ChatGPT for Math

George Grätzer

ChatGPT for Math

Springer

George Grätzer
Toronto, ON, Canada

ISBN 978-3-032-27285-0 ISBN 978-3-032-27286-7 (eBook)
https://doi.org/10.1007/978-3-032-27286-7

This Springer imprint is published by the registered company Springer Nature Switzerland AG
The registered company address is: Gewerbestrasse 11, 6330 Cham, Switzerland

ChatGPT is an* assistant, *not an oracle

Preface

My first real encounter with ChatGPT did not take place in a mathematics department. It happened in my study, prompted by simple curiosity. I wanted to see what this new tool could do in everyday settings. I experimented with writing emails, summarizing documents, translating text, drafting letters, and answering questions. The result of that curiosity was a book, *ChatGPT, Prompts and Beyond* (De Gruyter, 2025), in which I showed that ChatGPT could help with almost any task involving text. I even joked that it could do almost everything except cook your supper—though it would happily provide a recipe.

In that earlier book, mathematics played only a minor role. I treated ChatGPT primarily as a versatile writing tool, using it for editing, organizing prose, and exploring ideas, mostly through examples drawn from everyday life. I was not yet asking whether it might help in my professional world—the world of theorems, proofs, and LaTeX.

That question emerged gradually. Once ChatGPT became part of my routine, I began trying it on small mathematical tasks: polishing a paragraph, cleaning up the English in a referee report, and suggesting alternative phrasings for a definition. The results were good enough that I began to rely on them. The mathematics remained my responsibility, of course, but ChatGPT quickly proved to be excellent at improving prose.

The next step involved diagrams. Drawing ordered set diagrams in TikZ is satisfying but time-consuming. It requires positioning nodes, adjusting spacing, and correcting syntax errors, all of which demand patience. One day I described a small ordered set to ChatGPT and asked it to produce the corresponding TikZ diagram. To my surprise, the code not only compiled but produced a figure remarkably close to what I wanted. I asked for changes—different colors, more spacing, additional labels, a dotted edge—and each time the revised code was correct. Within an hour, I had a set of diagrams that would have taken much longer to construct by hand.

From there, I turned to LaTeX itself. Large projects inevitably produce long log files filled with cryptic error messages. Traditionally, diagnosing them required experience and persistence. On a whim, I began pasting fragments of log files into ChatGPT and asking, "What went wrong here? Please explain this in plain language." Often the answers were exactly what I needed. They pointed to a missing `\end{figure}`, an extra brace, a stray `\item`, or an incompatible package option. ChatGPT could not fix the file for me, but it greatly shortened the search.

By that point, I was using ChatGPT regularly for three kinds of mathematical work: drawing diagrams, improving mathematical prose, and diagnosing LaTeX issues. The natural next question was whether it could do more.

I began experimenting with proofs, cautiously. ChatGPT is a Large Language Model, not a proof assistant. It can produce arguments that sound convincing but are wrong. It can invent false theorems or overlook hypotheses. Still, it could rewrite proofs and suggest alternative strategies. When I gave it a proof and asked for a sentence-by-sentence critique, it often identified unclear steps and missing assumptions. I would never trust ChatGPT to certify correctness, but its help in clarifying and reorganizing arguments quickly became valuable.

From there, I moved on to examples and counterexamples. I asked for small probability spaces and random variables with particular properties: events that are pairwise independent but not mutually independent, random variables with identical distributions but different joint behavior, distributions with finite expectation but infinite variance, or processes exhibiting a specific dependence structure.

The answers varied. Some were wrong; many were surprisingly useful. Sometimes ChatGPT proposed an idea that I could refine. At other times, its attempt helped me articulate why the structure I requested could not exist. In this role, ChatGPT became a generator of candidate examples—imperfect, but often productive.

Meanwhile, I also tried using ChatGPT in my teaching. I asked it to design exercise sets, rewrite technical explanations for advanced undergraduates, draft short expository pieces for general readers, and simulate oral-exam questions. The pattern was consistent. ChatGPT rarely produced text that I could use unchanged, but it generated drafts and structures that I could easily adapt. Instead of facing a blank page when planning a lecture or an assignment, I could begin with a reasonably good outline.

At a certain point, it became clear that I was no longer merely "trying out" a new tool. I was using it consistently in my daily mathematical life—alongside LaTeX, TikZ, email, and the occasional computer algebra system. Not for everything—and never where exactness is essential. But whenever a task involved language, structure, or small computational experiments, I increasingly turned to ChatGPT.

This book grew out of that shift. It is not a theoretical study of artificial intelligence, nor a collection of isolated tricks. Instead, it presents workflows: concrete ways in which ChatGPT can help with proofs, examples, teaching, LaTeX, diagrams, bibliographies, indexing, and other aspects of mathematical work. Many of these workflows began as casual experiments—"Can ChatGPT help with this?"—and those included here are the ones I found myself returning to repeatedly.

Principle
Trust but verify. ChatGPT can suggest, rewrite, reorganize, and compute small cases. Responsibility for mathematical correctness rests entirely with the mathematician.

When used wisely, ChatGPT can make mathematical work more pleasant. It can lighten the burden of typesetting, suggest clearer explanations, reorganize a chapter, summarize a long computation, or produce a first—usually imperfect—version of an index. It cannot replace the moments of insight when a theorem comes together, but it can remove some of the surrounding clutter.

Remember, ChatGPT is an assistant, not an oracle. It produces text that must always be examined critically.

In my earlier book, I explored how a mathematician might use ChatGPT in everyday life. This book focuses on how to use it within mathematics. It addresses specific methods, expectations, and limitations. I hope the examples and patterns collected here help you integrate ChatGPT into your own practice—cautiously but confidently—and make at least some parts of your mathematical work easier and more enjoyable.

Toronto, ON, Canada
2026

George Grätzer

Introduction

As a mathematician, I have always relied on tools: I write with pencils and chalk, calculate with computers, draw diagrams with Adobe Illustrator, and prepare my papers and books with LaTeX. In recent years, a new tool has entered this landscape. Large Language Models, of which ChatGPT is the most visible example, now play a role. They do not reason as mathematicians do, yet they can produce, rewrite, and analyze mathematical text with surprising fluency. Used thoughtfully, they support many of the everyday tasks that shape mathematical work.

Principle

ChatGPT is an assistant, not an oracle.

It can draft, reorganize, rephrase, and suggest. But it cannot certify truth. Every mathematical statement it produces must be verified. This need for verification is a recurring theme of the book.

The chapters are organized into four parts, reflecting the different roles ChatGPT can play in mathematical practice.

Part I **(Chaps. 1–3): Foundations.** These chapters describe what ChatGPT can do in mathematical contexts. They introduce workflows for verifying its output. They establish the "assistant, not oracle" framework for the entire book.

Part II **(Chaps. 4–10): Core mathematical workflows.** This is the heart of the book. I show how ChatGPT can support small-case experimentation, help rewrite proofs, generate examples and counterexamples, assist with computations, and prepare diagrams. I also discuss uses in teaching and present prompt patterns tailored to mathematical work.

Part III **(Chaps. 11–16): Infrastructure and writing.** Mathematics involves more than theorems. It involves documents, notation, bibliographies, indices, diagrams, and exposition. Here I show how ChatGPT helps with consistency, reference organization, drafting, and LaTeX tasks.

Part IV **(Chaps. 17 and 18): Limits and outlook.** I discuss what ChatGPT cannot do and explain where caution is needed. I consider how these tools may shape mathematical writing in the future. The closing chapters offer reflections and a look ahead.

Appendix A collects prompt patterns for mathematicians in a compact, ready-to-use form.

Appendix B provides a concise guide to the LaTeX tools and workflows used throughout the book.

How to Read this Book You do not need to read this book strictly from beginning to end. Chapters 1–3 establish the conceptual framework used throughout: what ChatGPT can and cannot do, why verification is essential, and how to structure safe workflows. These chapters introduce terminology, principles, and habits that later chapters rely on, and I recommend that you read them first.

Beyond that, the book is deliberately modular. If you are primarily interested in proofs, examples, and counterexamples, you may focus on Chaps. 4–6. If your main interest is teaching, you may turn directly to Chaps. 9 and 15. If you are concerned with writing, diagrams, bibliographies, or LaTeX, you may consult Chaps. 8, 11–13, as needed.

Cross-references appear throughout the book, and some repetition is intentional: important principles reappear in different contexts. You will get the most value from this book by using it as a working companion—consulting specific chapters when a task arisesand returning to others over time—rather than by reading it as a strictly linear narrative.

This book grew out of my own everyday work with ChatGPT. I use it daily. Every chapter contains workflows I use in my research and writing. My hope is that this book helps other mathematicians make thoughtful use of these tools while preserving the precision, clarity, and independence of thought that define our discipline.

A Note on Standard Workflow A standard workflow is used throughout this book. I do not treat ChatGPT as a source of isolated answers. Instead, I use it within a small, repeatable workflow. The individual examples in later chapters may differ in detail—proofs, computations, diagrams, teaching materials—but the underlying pattern is remarkably stable. For ease of reference, I make that pattern explicit here.

A standard ChatGPT–math workflow

1. **Specify the mathematical goal precisely.** State exactly what you want: a definition, an example, a proof outline, a computation, a diagram, or a rewrite, together with any constraints on level, scope, or hypotheses.
2. **Generate a candidate.** Use ChatGPT to produce a first draft: text, example, computation, diagram, or code. Treat this output as provisional.
3. **Force explicit structure.** Ask for numbered steps, case splits, tables, or explicit computations. This often reveals hidden assumptions, missing cases, or unjustified inferences.
4. **Verify.** Check the result independently. This may involve small-case testing, direct computation, comparison with known results, or careful logical inspection. ChatGPT does not certify correctness.

5. **Rewrite, refine, or discard.** Accept what survives verification, revise what needs repair, and discard what fails. The final version is always the mathematician's responsibility.

This workflow embodies the central principle of the book: ChatGPT is an assistant, not an oracle. It accelerates drafting, exploration, and organization, but it never replaces mathematical judgment. In later chapters, I will frequently refer back to this pattern with brief remarks such as:

This follows the standard workflow introduced in the Introduction.

These references are intentional. They are meant to signal that what may look like a one-off example is in fact an instance of a general method. My aim is not to showcase clever prompts, but to teach a disciplined way of working—one that you can adapt to your own research, writing, and teaching with confidence.

On Durability and Scope Although this book refers throughout to ChatGPT, its central claims are not tied to any particular version or implementation. The workflow, failure modes, and limitations discussed here arise from structural features of large language models as a class: they generate plausible text rather than validated mathematics, lack semantic grounding, and have no intrinsic notion of truth. Specific examples will inevitably age, but the methodological principles—and the responsibilities that remain with the mathematician—are going to persist across model updates and future systems.

Contents

Part IV Limits and Outlook

Part I

Foundations

What Can ChatGPT Actually Do? 1

Principle
ChatGPT is an assistant, not an oracle.
It can reshape the *workflow* of mathematics, not the mathematics itself.

In this chapter, I survey several ways in which I use ChatGPT productively in mathematics. The emphasis is not on speculation. Instead, it focuses on concrete, day-to-day tasks that arise in my own work as a mathematician.

Mathematics is 10% inspiration and 90% perspiration.
Folklore

ChatGPT's role is to substantially reduce that 90%.

1.1 Supporting My Mathematical Research

A natural role for ChatGPT is to assist in exploring definitions and examples. When I work with a new concept, ChatGPT can point out hidden assumptions. It can also generate a variety of examples. For instance, if I introduce a new type of closure operator on a lattice, ChatGPT can test the definition on standard structures. It can also suggest variations. It can even generate minimal examples in which the operator behaves in unexpected ways.

Beyond generating examples, ChatGPT is particularly useful for giving *sketches* of proof strategies. For example, a complicated argument can be reformatted into a higher-level outline. In this way, the major steps are highlighted without committing to details. As a result, I can often spot gaps or inconsistencies before serious writing begins. Because ChatGPT can rewrite an argument at different levels

G. Grätzer, *ChatGPT for Math*, https://doi.org/10.1007/978-3-032-27286-7_1

of sophistication, I can easily produce an undergraduate-level explanation. I can also produce a research-level seminar explanation. I can even produce a refined proof outline from the same core idea.

Example

In practice, I use ChatGPT for the following tasks:

- supporting mathematical research through definitions, examples, and sketches of proofs;
- producing and refining diagrams in TikZ;
- assisting with LaTeX through rewriting and error diagnosis;
- computational and algorithmic assistance on small cases;
- teaching and communication tasks.

1.2 Producing Flow Charts

Flow charts provide a compact way to represent procedures in which actions branch, information is combined, and outcomes are reached step by step. They are widely used in algorithms and programming, business workflows, decision-making, education, engineering systems, and research methods. In each of these settings, the structure of the chart makes the underlying process explicit and transparent. ChatGPT generates TikZ code for flow charts fluently and can adjust layout, styling, coloring, and labeling to suit the intended interpretation.

As a simple example, consider the six-element flow chart F_6. Its elements are

$$\omega,\ \alpha,\ \beta,\ \gamma,\ \tau,\ \upsilon.$$

From the initial node ω, three branches lead to α, β, and γ. The two nodes β and γ then merge at τ, and the final node υ is reached from both α and τ. Thus the arrows of F_6 are: from ω to α, β, and γ; from β and γ to τ; and from α and τ to υ.

We obtain the diagram shown in Figure 1.1a, coded as:

```
\begin{tikzpicture}[>=stealth]

\node[vertex] (w) at (0,2) {};
\node[vertex] (a) at (-2,0.5) {};
\node[vertex] (b) at (0,0.5) {};
\node[vertex] (g) at (2,0.5) {};
\node[vertex] (t) at (1,-0.8) {};
\node[vertex] (u) at (0,-2) {};
```

```
\node[lab,above=2pt] at (w) {$\omega$};
\node[lab,left=2pt]  at (a) {$\alpha$};
\node[lab,left=2pt]  at (b) {$\beta$};
\node[lab,right=2pt] at (g) {$\gamma$};
\node[lab,right=2pt] at (t) {$\tau$};
\node[lab,right=2pt] at (u) {$\upsilon$};

\draw[->] (w)--(a);
\draw[->] (w)--(b);
\draw[->] (w)--(g);
\draw[->] (b)--(t);
\draw[->] (g)--(t);
\draw[->] (a)--(u);
\draw[->] (t)--(u);

\end{tikzpicture}
```

Figure 1.1a is not very nice. I then instruct ChatGPT that the two atoms on the right should be symmetric with respect to the line between the least element and the largest element and all the three coatoms be equidistant. See Figure 1.1b for the result.

Finally, I tell ChatGPT to add the labels and colors, obtaining Figure 1.1c.

The process is interactive. I can further request adjustments in spacing and node size. ChatGPT immediately rewrites the corresponding TikZ code.

This eliminates much of the trial-and-error cycle that traditionally makes diagram preparation slow and tedious. For more complex figures—commutative diagrams, automata, or hypergraphs—ChatGPT can generate initial layouts. I may then refine them; see Chapter 8. As an excellent exercise, look at the final product, Figure 1.1c. Try to draw it. You will soon see that four steps are not enough. You must determine the appropriate circle size and stroke thickness. You must also choose an appropriate stroke thickness for the edges...

1.3 Assisting with LaTeX and PDF

ChatGPT's proficiency in LaTeX is one of its strongest features for me. It can correct syntax errors. It can reorganize code. It can modernize outdated commands and identify problematic constructions. A particularly useful tool is *interactive editing*. ChatGPT rewrites the text sentence by sentence. It shows the revised version, gives a short explanation, and asks whether I want to accept the change; see Section 1.4. This allows me to maintain full control of the text while benefiting from expert-level editing assistance.

Equally valuable is ChatGPT's ability to work with a LaTeX log file. When a document fails to compile, the error messages are often cryptic. Even experienced users may struggle to locate the source of a problem. By copying a few lines

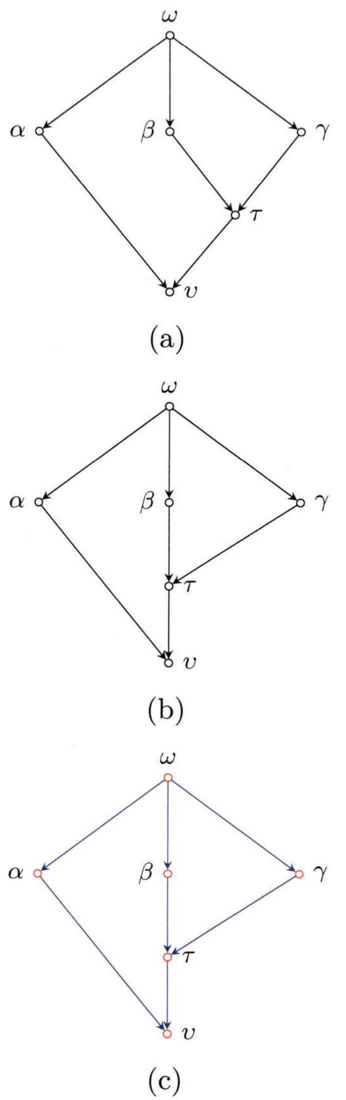

Fig. 1.1 Three drawings of the flow chart F_6

around the error into ChatGPT, I can usually obtain a clear and precise explanation. Suggested fixes often follow. Overfull boxes, missing braces, package conflicts, and undefined references can often be resolved in seconds. For more detail, see Section 16.3.

ChatGPT is also adept at generating or revising style files. I maintain custom classes for my books and lecture notes. When my book, *A Short History of the World* (Genesis Publishing, 2025), was completed, the publisher sent me a two-page-long specification for the final format. What a nightmare! Luckily, ChatGPT was there to do the hard work.

LaTeX produces PDFs. ChatGPT can do the reverse: it produces LaTeX code from a PDF. So there is no problem if you want to quote a result from a paper that you have only in PDF format. It can proofread a PDF for clarity, grammar, typos, and stylistic tone.

1.4 Editing Assistance

ChatGPT excels at editing. It can correct both grammar and style. It is really simple: use the prompt "edit:" and paste the text to be edited. ChatGPT returns the edited text.

It is easy, but not very useful. You get the edited text but no clue as to what changed and why. When comparing the edited text sentence by sentence, you may find changes you do not want. Tedious work.

The deluxe version of editing is *interactive editing*. You edit sentence by sentence, ChatGPT displays your **original sentence**, the **corrected version**, and the **reason for the change**.

Then there is a prompt

`Accept change? (y/n):`

If I type `y`, ChatGPT permanently applies the editing; if I type `n`, the original sentence is unchanged.

At the end, I get the corrected text with the changes **I approved**.

Remember: ChatGPT is a predictive language model. Given a sequence of words, it produces the word that is statistically likely to follow. So it is best suited for editing.

1.5 Computational and Algorithmic Assistance

Mathematics is an experimental science. We once carried out experiments by pencil and paper. Now ChatGPT allows us to perform experiments on a much larger scale.

Such computations are valuable when I explore conjectures and test ideas on small cases. When studying whether a property is preserved by a construction, ChatGPT can rapidly generate examples. It can run through cases and summarize findings. Although I must verify the results, this exploratory feedback often reveals patterns or counterexamples that would be difficult to find manually. See Chapter 4 for more detail.

1.6 Teaching and Communication

For teaching, ChatGPT provides an array of tools that I rely on frequently. It can generate exercises of varying difficulty. It can create quizzes with well-designed distractors. It can also produce detailed solutions. When preparing lectures, ChatGPT can summarize a topic. It can give intuitive explanations. It can create illustrative examples or analogies. It can adapt explanations for different audiences. It can adapt explanations for different audiences, including undergraduate students, graduate students, and researchers. See Chapters 9 and 15.

For writing papers and books, ChatGPT excels at improving clarity. Rough paragraphs can be rewritten into crisp, well-structured prose. Long proofs can be reorganized into lemmas and propositions. The ability to switch seamlessly between informal explanation and formal statement is especially helpful to me.

1.7 Transition

The examples in this chapter show what ChatGPT can do in mathematical practice. To use it effectively, however, it is not enough to see its capabilities—we must also understand how it produces its output.

In the next chapter, I examine ChatGPT as a predictive language model. I describe the mechanisms behind its responses, its characteristic strengths, and its typical failure modes. This perspective is essential for interpreting its output and for using it safely in mathematical work.

How ChatGPT Interacts with Math

2

ChatGPT is not a theorem prover, a computer algebra system, nor a database of mathematical results. It is a predictive language model. Given a sequence of words (tokens), it produces the word (token) that is statistically likely to follow. For my mathematical work, understanding this simple fact is essential for using the system effectively and safely.

In this chapter, I examine how ChatGPT "thinks" from a mathematical point of view. I describe its strengths and its characteristic failure modes, and I give a practical framework for interpreting its output. My goal is not to give a technical description of Large Language Models (LLMs). It is to provide myself—and other working mathematicians—with a reliable mental model of what the system can and cannot do.

2.1 ChatGPT as a Predictive Text Model

ChatGPT has been trained on a large corpus of text, including expositions, proofs, tables, code, and informal discussions. Its task is to continue a passage in a way that resembles the patterns it has seen. It does not maintain an internal stock of verified theorems or a library of facts, and it does not compute formal consequences of axioms. When it produces a proof, it is generating text that *looks like* a proof.

For example, if most textbook proofs of a certain inequality begin by introducing an arbitrary element and expanding a square, ChatGPT may do the same whether or not it is necessary. If most discussions of a concept invoke a standard lemma, it may appeal to that lemma even when its hypotheses do not apply.

The system does not aim at correctness; **it aims at plausibility**.

G. Grätzer, *ChatGPT for Math*, https://doi.org/10.1007/978-3-032-27286-7_2

2.2 Strengths Relevant to My Work

Despite these limitations, ChatGPT has significant strengths that can support my mathematical work.

► **Strengths and Failure Modes at a Glance**

Strengths	Failure modes
Mastery of mathematical form	Hallucinated theorems and references
Pattern recognition across examples	Plausible but incorrect proofs
Flexible rewriting at multiple levels	Misuse or drift in terminology

2.2.1 Mastery of Form

ChatGPT has absorbed the stylistic conventions of mathematical writing: the structure of definitions, the layout of proofs, the rhythm of exposition, and the conventions of LaTeX. As a result, it can:

- rewrite draft text in clearer language,
- adapt exposition to different audiences,
- generate outlines and structured summaries,
- translate informal notes into formal prose,
- produce compilable LaTeX for common tasks.

These abilities make it a useful assistant for both teaching and writing.

Worked Example: From Informal Notes to Polished Exposition

2.2.2 Raw Notes

Let V be a finite-dimensional vector space. Every vector in V can be written as a linear combination of basis vectors contained in a fixed basis.
Proof: straightforward? If v is not already a basis vector, write it as a linear combination of other vectors spanning V, and repeat. The process terminates since the space is finite-dimensional.

2.2.3 Prompt to ChatGPT

Rewrite the following notes as a short textbook-style proposition with a clear statement and proof. Assume the reader knows basic ring theory but not this argument. Keep it concise and use standard LaTeX structure.

2.2.4 ChatGPT's Rewritten Version

Proposition 2.1 *Let $(\Omega, \mathcal{F}, \mathbb{P})$ be a finite probability space. Every event $A \in \mathcal{F}$ can be expressed as a union of atoms contained in A.*

Proof Let $A \in \mathcal{F}$. If A is an atom, there is nothing to prove. Otherwise, A contains a proper nonempty subset $B \in \mathcal{F}$. If B is not an atom, apply the same decomposition to B. Since the probability space is finite, this process terminates after finitely many steps, expressing A as a union of atoms contained in A. □

2.2.5 What This Illustrates

- Informal reasoning is converted into standard mathematical form.
- The argument is organized into a clear statement-proof structure.
- Missing justifications are supplied at an appropriate level.
- The result is immediately compilable LaTeX.

2.2.6 Pattern Recognition

Because ChatGPT has seen countless examples of proof techniques, it is good at suggesting next steps in a familiar style:

- introducing a key element,
- identifying a likely lemma,
- splitting into cases,
- recognizing standard counterexamples.

While these suggestions must be checked, they often guide me toward useful lines of thought.

Worked Example: Recognizing a Proof Pattern

Problem (as Stated) Let $\mathcal{K}$ be an integral operator on a finite interval $[a, b]$ with kernel $K(x, t)$. Assume that for every subinterval of $[a, b]$ of length 2, the restriction of $\mathcal{K}$ to that subinterval is linear and bounded. Prove that $\mathcal{K}$ is linear and bounded on the entire interval $[a, b]$.

Initial Thoughts The hypothesis has a clear local-to-global flavor. If the conclusion were false, one would expect a minimal interval on which linearity or boundedness fails. Such a minimal obstruction should already appear on a short subinterval, suggesting a proof by contradiction that isolates a local defect and propagates it.

Prompt to ChatGPT

> Suggest a proof strategy for the following statement from probability theory. Do not give a complete proof; instead, identify the likely next steps and the standard lemmas or principles that are typically used in this setting.

ChatGPT's Response (Strategy Outline)

- Assume, for contradiction, that the statement fails and fix a minimal counterexample (for example, a smallest σ-algebra, index set, or time interval on which the conclusion does not hold).
- Identify the relevant local property (such as finite additivity, independence on generating events, or boundedness in expectation) that is assumed to hold on simpler or smaller structures.
- Use a standard extension principle—such as the π–λ theorem, the monotone class theorem, or a limiting argument based on dominated convergence—to propagate the local property to the larger structure.
- Show that the assumed counterexample violates this extension principle, contradicting minimality.

How This Helps The response does not produce a proof, but it correctly identifies a familiar probabilistic proof pattern:

- reduction to a minimal or extremal counterexample,
- appeal to a standard structural theorem for extending local properties,
- passage from finite or local behavior to a global conclusion.

This is precisely the line of attack an experienced probabilist would try first. Each step still requires careful verification, but recognizing the pattern narrows the search space and points toward the right technical tools.

2.3 Characteristic Failure Modes

To use ChatGPT safely, I must recognize its typical mistakes. Three failure modes occur repeatedly.

2.3.1 Hallucinated Theorems and References

ChatGPT may confidently cite a "well-known" theorem that does not exist, attribute a result to the wrong author, or describe a lemma with incorrect hypotheses. This behavior is not deception; it is a natural consequence of predicting plausible continuations.

2.3.2 Plausible but Incorrect Proofs

Many generated proofs follow standard rhetorical patterns but contain subtle logical errors. A hypothesis may be used without being stated, a forward implication may be applied in reverse, or a diagram may be drawn for a structure that does not satisfy the required axioms.

2.3.3 Misuse of Terminology

Occasionally the system uses mathematical terms imprecisely: "join" in a semilattice, "basis" for a spanning set, or "continuous" where monotone is intended. These slips are small but significant.

► **Warning**
ChatGPT can produce arguments that look like correct mathematical proofs while hiding subtle errors in logic or hypotheses.
Every proof still requires human verification.

2.4 A Practical Framework: Assistant, Not Oracle

The central principle I follow throughout this book is:

Principle
Treat ChatGPT as an assistant, not as an oracle.

Used as an assistant, ChatGPT can generate examples, reorganize arguments, propose alternative viewpoints, and help expose hidden assumptions. Used as an oracle—to certify correctness or settle a conjecture—it is unreliable. A language model cannot guarantee truth.

2.4.1 A Diagnostic LaTeX Example

Example

Goal
Identify the real source of a cryptic LaTeX error message and apply the smallest reliable fix.

(continued)

Error reported

```
Extra }, or~forgotten \endgroup.
\@endfloatbox ...\egroup \color@endbox
l.206 \end{figure}
```

The error is reported at `\end{figure}`, which is almost never where the real problem lies.

Faulty caption

```
\caption{The subgroup lattice generated
by the set $\{a,b,c$,
where $a<b$ and $c<d$}
```

Diagnosis
The set notation `\{a,b,c` opens a group that is never closed. As a result, LaTeX reaches `\end{figure}` while still inside a group, triggering the reported error.

Corrected caption

```
\caption{The subgroup lattice generated
by the set $\{a,b,c\}$,
where $a<b$ and $c<d$}
```

The diagnosis and the fix are mine. ChatGPT's role is to interpret the log message and suggest likely failure modes. Verification remains my responsibility.

2.5 General Guidelines for My Mathematical Practice

From experience, several guidelines help ensure productive interactions.

2.5.1 Be Specific

Vague prompts lead to vague answers. A request such as "Explain this theorem" produces generic text. A request such as "Rewrite this proof for a strong undergraduate and separate it into numbered steps" produces something genuinely useful.

2.5.2 Work Interactively

ChatGPT improves with context. I ask follow-up questions. I request alternatives. I refine specifications. I correct mistakes. The dialogue matters as much as the initial prompt.

2.5.3 Check Everything

No statement should be accepted without verification. When ChatGPT provides an example, I must verify it. When it produces a proof, I must inspect it. The responsibility for correctness remains with me.

2.5.4 Example: A Short Explanatory Paragraph

I will illustrate these guidelines with a small but typical task. I wanted a short paragraph explaining why every finite Boolean algebra is isomorphic to the power set of a finite set.

This example illustrates the standard workflow introduced in the Introduction, applied to a short explanatory task.

Example

Goal

A three-to-five sentence explanation, suitable for advanced undergraduates, with no proofs but a clear indication of why the statement is true.

Workflow

- *State a precise goal.* I began by writing down what I wanted: a short explanatory paragraph about finite Boolean algebras, positioned between their definition and Birkhoff's representation theorem.
- *Give context and constraints.* I told ChatGPT that the paragraph would appear in a chapter on lattices, after the definition of Boolean algebras and before the representation theorem, and that finite sets and power sets had already been introduced. I also specified that the answer should be a single paragraph of prose, with no bullet points and no formulas beyond basic notation such as 2^n.
- *First draft* (*ChatGPT*) ChatGPT produced the following paragraph:

  > A finite Boolean algebra has a very rigid structure. Its elements can be understood in terms of simpler building blocks, called atoms, which play a role analogous to singletons in a power set. This analogy helps explain why finite Boolean algebras resemble power sets and why their structure is so well behaved.

  The paragraph was readable, but it avoided the crucial mathematical point.
- *Correction prompt.* Since the key fact is that *every element is a join of atoms*, I issued a targeted revision request:

  > Please rewrite the paragraph and explicitly state that every element of a finite Boolean algebra is a join of atoms, and that atoms correspond to singletons in the associated power set.

Example

- *Second draft (ChatGPT).* ChatGPT responded with the following revised paragraph:

  > A finite Boolean algebra has a very rigid structure. Every element can be written as a join of atoms, the minimal nonzero elements of the algebra. These atoms correspond naturally to singletons in a power set, which explains why finite Boolean algebras closely resemble power sets and why their behavior is so well controlled.

- *Verification and final editing.* The revised paragraph was mathematically correct. I shortened one sentence, adjusted terminology to match the rest of the chapter, and added a forward reference to the later section where the representation theorem is proved.

The final paragraph in the book is mine. I verified the mathematics, chose the emphasis, and edited the prose. ChatGPT's role was to generate an initial draft quickly and to respond accurately to precise correction prompts.

2.5.5 Another Example: A Cryptic LaTeX Error Message

Example

Goal
Identify the real cause of a cryptic LaTeX error message, locate it precisely, and apply the smallest reliable fix.

Workflow

- *The error reported by LaTeX.* During compilation, LaTeX stopped with the following message:

```
! Extra }, or~forgotten \endgroup.
\@endfloatbox ...\egroup \color@endbox
l.206 \end{figure}
```

(continued)

The error was reported at `\end{figure}`, which is almost never where the real problem is.

- *The problematic caption.* The figure contained the following caption:

```
\caption{The lattice $L$ generated
by $\{a,b,c\}$, where $a<b$ and $c<d$}
```

At first glance, this looks harmless.

- *First diagnosis prompt.* I paste the log fragment into ChatGPT and ask:

  Explain this error message in plain language. Assume I am working inside a figure environment. Where should I start looking?

Example

- *First diagnosis.* ChatGPT replies:

  This error usually means that there is an extra closing brace } or a missing opening brace earlier in the figure environment. The problem is not at `\end{figure}` itself, but somewhere inside the figure, often in the caption or in a command argument.

  This narrowed the search but did not yet identify the cause.

- *Refinement prompt.* I then ask:

  Be more specific. Look carefully at the caption and explain what might be wrong.

- *Refined diagnosis.* ChatGPT responds:

  In the caption, the set $\{a, b, c\}$ opens with \{ but is never properly closed. The closing brace } at the end of the caption matches the argument of `\caption`, not the set. This causes LaTeX to reach `\end{figure}` while still inside a group.

- *Corrected caption.* The fix was to close the set explicitly:

```
\caption{The lattice $L$ generated by
$\{a,b,c\}$, where $a<b$ and $c<d$}

\end{itemize}
```

(continued)

After adding the missing brace, the document compiled without error.

The diagnosis and the fix are mine. I verified the correction by recompiling. ChatGPT's role was to interpret the log message, inspect the caption carefully, and articulate why the error surfaced far from its true cause.

Used in this way, and following the guidelines above, ChatGPT is a helpful assistant rather than an oracle.

2.6 Exercises

Exercise 2.1 Describe in words a lattice or ordered set diagram you would like to have. For example, you might specify "a modular lattice of length 4 with exactly three coatoms". Ask ChatGPT to propose TikZ code for the diagram. Decide which parts you would keep and which parts you would edit by hand.

Exercise 2.2 Find a warning or error message in a LaTeX log files that refers to a line number far from the actual problem. Ask ChatGPT to explain why LaTeX reports the error where it does and to suggest what kinds of source constructs typically cause such behavior.

Exercise 2.3 Choose a mathematical notion you know well (for example, group, eigenvalue, or continuity). Ask ChatGPT to explain it at three levels: (a) advanced undergraduate, (b) beginning graduate student, (c) research mathematician. Compare the explanations. Which features change and which remain invariant?

Exercise 2.4 Ask ChatGPT to state a nontrivial theorem in an area you know reasonably well. Without consulting external sources, inspect the statement for warning signs: imprecise hypotheses, missing conditions, or overly strong conclusions. Rewrite the prompt to reduce the risk of hallucination.

Exercise 2.5 Ask ChatGPT the same mathematical question twice, changing only one phrase in the prompt (for example, replacing "explain" by "give a precise definition"). Compare the responses. What does this reveal about how ChatGPT interprets intent?

Exercise 2.6 Write a short informal mathematical paragraph of your own. Ask ChatGPT to paraphrase it without changing the mathematics. Then ask ChatGPT to paraphrase it for clarity. Compare the results. Which mathematical information is most fragile under paraphrasing?

Exercise 2.7 Ask ChatGPT to solve a routine mathematical problem. Then ask it explicitly to verify each step of its own solution. Identify any step that was revised, weakened, or corrected. What role did verification play in improving reliability?

Exercise 2.8 Take a loosely structured mathematical explanation (for example, from lecture notes or a preprint). Ask ChatGPT to impose structure by introducing definitions, lemmas, and a proof outline. Decide which structural changes genuinely improved the exposition.

Exercise 2.9 Ask ChatGPT a question that is ambiguous or underspecified. Observe how it resolves the ambiguity. Rewrite the prompt to remove the ambiguity and compare the responses. What assumptions does ChatGPT make when forced to guess?

2.7 Transition

In this chapter, I have examined how ChatGPT produces its output and why that output must be treated with caution. Understanding the system as a predictive language model explains both its strengths and its characteristic failures.

But understanding alone is not enough.

To use ChatGPT effectively in mathematics, we must develop disciplined habits for checking what it produces. Every definition, example, and proof must be verified, not merely read.

In the next chapter, I turn to the central practice that makes ChatGPT usable in mathematics: verification. I present concrete workflows for testing examples, analyzing proofs, and locating errors. These methods transform ChatGPT from a source of plausible text into a reliable assistant.

Verification: Trust but Verify

3

Principle
ChatGPT can draft proofs and construct examples, but *every* mathematical claim it produces still needs verification.

The previous chapter explained *why* verification is essential. Here we focus entirely on how to do it, step by step.

All of the verification techniques in this chapter are instances of the standard workflow presented in the Introduction, with particular emphasis on the verification step.

ChatGPT can draft proofs. It can construct examples. It can suggest counter-examples, but it is not a formal reasoning engine. It writes mathematical prose that *looks* correct. Sometimes it is correct and sometimes it is almost correct. To use ChatGPT effectively, we must develop habits of verification. In this chapter, I describe practical workflows that mathematicians can adopt when checking the correctness of text produced by a language model.

3.1 Why Verification Matters

A language model predicts the next token. It does not manipulate algebraic structures. It does not check that a map is well defined. It does not confirm that a lemma follows from its hypotheses. The result is that ChatGPT often gives:

- arguments with a missing hypothesis,
- incorrect examples that violate a hidden condition,

G. Grätzer, *ChatGPT for Math*, https://doi.org/10.1007/978-3-032-27286-7_3

- proofs that use a statement proved later in the text,
- or explanations that are true for groups but false for semigroups.

None of this is surprising. ChatGPT is an assistant, not an oracle. But with the right workflow, verification becomes fast and reliable.

▶ Checking a ChatGPT Proof

1. Restate the statement yourself and compare with ChatGPT's version.
2. Ask ChatGPT to outline the proof in numbered steps.
3. Test the statement on small finite structures when possible.
4. Look for hidden assumptionsindexHidden assumptions or missing cases.
5. If possible, request an alternative proof and compare.

Example

Worked example: checking a ChatGPT proof
Statement If a function $f : [a, b] \to \mathbb{R}$ is differentiable on (a, b) and $f'(x) = 0$ for all $x \in (a, b)$, then f is constant on $[a, b]$.

Step 1: Restate the statement. I rewrote the claim for myself as:

> A differentiable function with identically zero derivative cannot change value: for all $x, y \in [a, b]$, we must have $f(x) = f(y)$.

This matches ChatGPT's formulation.

Step 2: Ask for a proof outline. I asked ChatGPT:

> Outline a proof of this statement in numbered steps. Do not assume any results that have not been stated explicitly.

ChatGPT's proof outline.

1. If $f'(x) = 0$, then f has slope 0 everywhere.
2. A function with slope 0 everywhere cannot increase or decrease.
3. Therefore, f is constant.

At first glance, this looks plausible, but it is too informal.

Step 3: Test on simple examples. I tested the claim on familiar differentiable functions: constant functions (true), $f(x) = x$ (derivative not 0), and $f(x) = x^2$ (derivative not 0). The statement behaves as expected. This supports the claim but does not justify the outline.

(continued)

Step 4: Look for hidden assumptions. The outline relies on an unstated bridge: it treats "slope everywhere" as if it directly forces global behavior. But the precise link between local derivative information and global equality is a theorem. Without naming it, the outline is not a proof.
Step 5: Request an explicit proof with the key theorem. I asked ChatGPT:

> Give a proof that explicitly uses a standard theorem (e.g., the Mean Value Theorem). State the theorem you use and show exactly where it is applied.

Second proof attempt (ChatGPT). Let $x, y \in [a, b]$ with $x < y$. Since f is differentiable on (a, b), it is continuous on $[a, b]$, so the Mean Value Theorem applies on $[x, y]$. Therefore, there exists $c \in (x, y)$ such that

$$f'(c) = \frac{f(y) - f(x)}{y - x}.$$

But $f'(c) = 0$ by assumption, hence $f(y) - f(x) = 0$, so $f(y) = f(x)$. Since x and y were arbitrary, f is constant on $[a, b]$.

Verification I checked the hypotheses needed for the Mean Value Theorem: continuity on $[x, y]$ and differentiability on (x, y). Differentiability was assumed, and it implies continuity. The conclusion $f(y) = f(x)$ follows by straightforward algebra. This time, the argument is complete.

The final proof is mine to accept or reject. The first outline used intuitive language ("slope") without a theorem. Only after forcing an explicit theorem and checking its hypotheses did I accept the argument. ChatGPT's role was to propose candidates; verification remained my responsibility.

3.2 Asking ChatGPT to Critique Itself

One effective technique is to ask ChatGPT to act as its own referee.

Example

Prompt

> Here is a proof you wrote. Critique it. Look for gaps, missing cases, incorrect implications, and dependence on unproved results. Do not attempt to repair the proof; only analyze it.

(continued)

Worked example

Statement If $f : [a, b] \to \mathbb{R}$ is differentiable on (a, b) and $f'(x) = 0$ for all $x \in (a, b)$, then f is constant on $[a, b]$.
Proof (as written). Assume $f'(x) = 0$ for all $x \in (a, b)$. Then the slope of f is 0 everywhere, so the function cannot increase or decrease. Therefore f is constant on $[a, b]$.

Critique request. I submitted the proof above to ChatGPT using the prompt.

Example

ChatGPT's critique.

> The proof is informal and leaves the main implication unjustified. The statement "slope 0 everywhere" is an intuition, not a theorem. To conclude that f cannot increase or decrease, one needs a result that connects the derivative to global behavior on an interval.
> In particular, the proof never cites or verifies the hypotheses of a theorem such as the Mean Value Theorem. Without invoking such a result, the argument has a gap: it does not show that for arbitrary $x < y$ one must have $f(x) = f(y)$.
> A second issue is that the conclusion claims constancy on $[a, b]$ while the hypothesis only states differentiability on (a, b). This is not necessarily a problem, but the proof does not explain why values at the endpoints are controlled (for example, by continuity implied by differentiability on the open interval together with an extension argument).
> As written, the proof reads as a heuristic explanation rather than a complete proof: it depends on an unstated theorem and does not check its required assumptions.

This critique is correct. The proof replaces a theorem with intuition and never makes its dependencies explicit. By asking for criticism rather than repair, I forced ChatGPT to identify the missing bridge between local information (the derivative) and the global conclusion (constancy). This role-switching is especially effective in real analysis, where short, plausible arguments often hide theorems that do the real work.

I usually follow up by asking it to mark any step that uses an unstated hypothesis or a lemma that has not yet been proved. Even if the analysis is imperfect, it helps localize the weak points.

3.3 Redundant Prompting

A second technique is to request multiple, independent versions of the same argument.

Example

Prompt

Give me a second proof of this lemma, using a different method.

Worked example

Lemma. Let $f: [a, b] \to \mathbb{R}$ be differentiable on (a, b). If $f'(x) = 0$ for all $x \in (a, b)$, then f is constant on $[a, b]$.

First proof (via the Mean Value Theorem). Let $x, y \in [a, b]$ with $x < y$. Since f is differentiable on (a, b), it is continuous on $[a, b]$, so the Mean Value Theorem applies on $[x, y]$. Thus there exists $c \in (x, y)$ such that

$$f'(c) = \frac{f(y) - f(x)}{y - x}.$$

Because $f'(c) = 0$, we obtain $f(y) = f(x)$. Since x and y were arbitrary, f is constant on $[a, b]$.

Second proof request. I asked ChatGPT:

Give me a second proof of this lemma using a different method.

Second proof (monotonicity argument) (ChatGPT). Fix $x < y$ in $[a, b]$. For every $t \in (x, y)$, we have $f'(t) = 0$. In particular, $f'(t) \geq 0$ for all $t \in (x, y)$, so f is nondecreasing on $[x, y]$. Likewise, $f'(t) \leq 0$ for all $t \in (x, y)$, so f is nonincreasing on $[x, y]$. Hence f is both nondecreasing and nonincreasing on $[x, y]$, which implies that f is constant on $[x, y]$. Since x and y were arbitrary, f is constant on $[a, b]$.

I compared the two proofs. They do not share the same surface structure: the first reduces the statement to a single application of the Mean Value Theorem, while the second factors the conclusion through monotonicity. But they rely on the same substantive dependency: the theorem connecting the sign of the derivative to global behavior. Because both arguments ultimately rest on standard results of real analysis, and no extra hypotheses have been introduced, I gain confidence that the lemma is being used correctly rather than accidentally.

3.4 Small-Case Testing

For many algebraic or combinatorial questions, verification starts with small cases. ChatGPT can enumerate:

- all groups of order 8,
- all partitions of a set of size 4,
- all sublattices of a finite lattice,
- all homomorphisms between two small finite structures.

Once the small cases are explicit, we can test conjectures mechanically.

Example

Prompt

List all groups of order 8 up to isomorphism, and for each one test whether property (P) holds.

ChatGPT can supply the list. It can even write small Python or SageMath functions to test associativity, distributivity, or other laws. I then run and check the code myself.

Working at small cardinalities often reveals counterexamples quickly, or shows that a statement is true for trivial reasons in low dimensions but fails in general.

3.5 Localizing Errors

When a proof seems incorrect, it is often not necessary to discard it entirely. More often, a single step is unjustified, misstated, or silently uses an unstated hypothesis. The goal is to find that step.

▶ Localizing the Weak Step

1. Ask ChatGPT to rewrite the proof as a numbered list of steps.
2. For each step, ask whether the inference is logically valid.
3. Check whether all hypotheses used at that step have been stated.
4. Mark steps that rely on an undefined term or a future result.

Example

Worked example

Claim

If events A and B are independent, then A is independent of B^c.

Proof (as written).

Assume A and B are independent, so $P(A \cap B) = P(A)P(B)$. Since

$$A = (A \cap B) \cup (A \cap B^c),$$

we have

$$P(A) = P(A \cap B) + P(A \cap B^c).$$

Substituting $P(A \cap B) = P(A)P(B)$ gives

$$P(A \cap B^c) = P(A) - P(A)P(B) = P(A)P(B^c).$$

Therefore A is independent of B^c.

The argument is plausible, but something feels too fast.

Step 1: Numbered rewrite. I asked ChatGPT to rewrite the proof as explicit steps. It produced:

1. Assume A and B are independent.
2. Then $P(A \cap B) = P(A)P(B)$.
3. Use the decomposition $A = (A \cap B) \cup (A \cap B^c)$ to write $P(A) = P(A \cap B) + P(A \cap B^c)$.
4. Substitute $P(A \cap B) = P(A)P(B)$ and solve for $P(A \cap B^c)$.
5. Conclude that $P(A \cap B^c) = P(A)P(B^c)$.
6. Therefore A is independent of B^c.

Step 2: Check each inference. Steps 1 and 2 are the definition of independence. Step 3 uses a set identity; it is correct only if the union is disjoint. Step 4 is algebra. Step 6 is the definition of independence again.

Step 3: Isolate the weak step. Step 3 is the critical claim:

The sets $A \cap B$ and $A \cap B^c$ form a disjoint union equal to A.

The equality is true, but the disjointness is what justifies replacing

$$P\big((A \cap B) \cup (A \cap B^c)\big)$$

by a sum of probabilities.

(continued)

Step 4: Check hypotheses and dependencies. The proof implicitly uses a basic probability rule: if E and F are disjoint events, then $P(E \cup F) = P(E)+P(F)$. That rule must be available (and the disjointness must be stated or checked). Otherwise the step from a set decomposition to an additive formula is not valid.

Conclusion. The proof is essentially correct, but it is missing a required justification at Step 3: it must note that $(A \cap B) \cap (A \cap B^c) = \varnothing$. Without that, the key probability identity is used without its hypothesis.

ChatGPT did not tell me whether the proof "felt right". Instead, by forcing a step-by-step rewrite, it made the hidden dependency impossible to ignore. The localization of the issue reduced a vague sense of unease to a precise question: *Have we actually checked the hypothesis needed to add these probabilities?*

Once the weak step is isolated, the remaining work becomes sharply focused and much more efficient. "ChatGPT as a proof checker" works best when it is forced to expose the logical joints rather than to smooth them over.

3.6 Checking Examples and Counterexamples

Verification is especially important for examples. If ChatGPT claims:

This is a nondistributive lattice.

I ask it to carry out the critical computations.

Example

Prompt

For runs in Bernoulli trials, compute the probability of

at least one run of length 2,

a run starting at position i or $i + 1$,

and compare the results. State explicitly where naive additivity fails.

Worked example

Context. Consider a sequence of n independent Bernoulli trials with success probability p. Let R_i denote the event that a run of two successes occurs starting at position i (that is, trials i and $i + 1$ are both successes).

A common but dangerous intuition is to treat these events as if they were disjoint or independent.

(continued)

Naive claim (as written).

$$P(\text{at least one run of length 2}) = \sum_{i=1}^{n-1} P(R_i).$$

The claim is plausible but suspicious.

Step 1: Compute the individual terms. For each i,

$$P(R_i) = p^2.$$

Thus the right-hand side equals $(n-1)p^2$.
Step 2: Check overlap between runs. For R_i and R_{i+1}, both occur exactly when trials $i, i+1, i+2$ are all successes. Hence

$$P(R_i \cap R_{i+1}) = p^3.$$

This overlap is the critical case.
Step 3: Compute a concrete run. Take $n = 3$. Then

$$P(\text{at least one run of length 2}) = P(R_1 \cup R_2).$$

By inclusion–exclusion,

$$P(R_1 \cup R_2) = P(R_1) + P(R_2) - P(R_1 \cap R_2) = 2p^2 - p^3.$$

Step 4: Compare with the naive sum. The naive sum gives $2p^2$, while the correct probability is $2p^2 - p^3$. The difference is exactly p^3, the probability of a run of three successes.

Conclusion. The naive additivity fails precisely because overlapping runs can occur simultaneously. The missing correction term corresponds to longer runs, which are invisible if one only counts runs locally.

This run-by-run computation exposes the hidden dependency. Just as in algebraic distributivity, the failure is not mysterious: it comes from overlap. By forcing explicit calculations on minimal configurations, the precise source of the error becomes clear, and further corrections (via inclusion–exclusion or recursion) can be introduced systematically.

The same method works for many algebraic structures. This includes semigroups, ordered sets, rings, and Boolean algebras. The key is to ask for explicit computations, not just verbal assurances.

3.7 Verifying Definitions and Hypotheses

ChatGPT often supplies the right definition, but with a missing hypothesis or a slightly nonstandard formulation. A reliable workflow is:

► **Stress-Testing a Definition**

1. Ask for the definition.
2. Ask for three examples.
3. Ask for three nonexamples.
4. Ask which hypotheses are actually used in a standard theorem.

If the examples or nonexamples contradict the definition, the definition may be incorrect or incomplete.

Example

Diagnosing a misleading LaTeX error
The error reported.

```
./ChatGPTMath.tex:2409: Undefined control
sequence. \endtcolorbox ->\tcb@insert
@after@part ... l.2409 \end{examplebox}
```

The error is reported at `\end{examplebox}`.

What the error actually means. The undefined control sequence is `\endtcolorbox`, an internal command provided by the `tcolorbox` package. The error therefore does *not* originate in the content of the `examplebox` environment nor in its closing command. It indicates that `tcolorbox` has been asked to close a box without having all of its required internal definitions available.

Likely cause. The `examplebox` environment is defined using `tcolorbox` features that require additional libraries (such as `breakable` or `skins`), but these libraries have not been loaded. As a result, LaTeX reaches `\end{examplebox}` and expands `\endtcolorbox`, only to discover that some internal macros are undefined.

Why the location is misleading. The error is reported at the end of the environment because this is where LaTeX attempts to execute the internal box-closing code. The actual problem was established earlier, when the environment was entered without all required definitions in place. This pattern is typical for errors involving complex environments.

Fix. Load the required `tcolorbox` libraries in the preamble:

(continued)

```
\usepackage{tcolorbox}
\tcbuselibrary{breakable,skins,hooks}
```

Irrelevant warning. The message

```
Underfull \hbox (badness ...)
```

is a typographic warning and is unrelated to this error.

This example illustrates a common LaTeX principle: the reported location of an error often marks the point where processing fails, not where the error was introduced. Recognizing this pattern turns a cryptic message into a routine fix.

3.8 The Human in the Loop

ChatGPT excels at fast drafting and fast revising, but verification remains a human responsibility. With good prompting, however, the process becomes lighter.

► **Best Practice**

Before I accept any ChatGPT-generated argument, I check:

- that the statement is correctly formulated,
- that every inference is justified,
- that small cases and edge cases agree,
- that alternative formulations do not reveal contradictions.

A typical meta-prompt is:

Example

Prompt

Here is an argument. List the weak points: any claims that need justification, any hidden hypotheses, and any steps that might fail in a noncommutative or infinite setting.

Worked example

Claim. Let $f\colon [a,b] \to \mathbb{R}$ be differentiable. If $f'(x) = 0$ for all $x \in (a,b)$, then f is constant on $[a,b]$.

(continued)

Argument (as written).

Assume $f'(x) = 0$ for all $x \in (a, b)$. Then the slope of f is zero everywhere. A function with zero slope cannot increase or decrease. Therefore f is constant on $[a, b]$.

Weak-point analysis request. I submitted the argument above to ChatGPT using the prompt.

ChatGPT's analysis.

1. The argument relies on an intuitive notion of "slope", but does not state or invoke any theorem that connects the derivative to global behavior. This is a missing justification.
2. The conclusion that a function with zero derivative cannot increase or decrease implicitly assumes a result such as the Mean Value Theorem or a monotonicity theorem, neither of which is stated.
3. The claim concerns the closed interval $[a, b]$, but the hypothesis only assumes differentiability on (a, b). The argument does not explain why values at the endpoints are controlled.
4. The argument does not address whether additional hypotheses (such as continuity on $[a, b]$) are required, nor does it note that differentiability implies continuity.
5. The reasoning is specific to real-valued functions on intervals. In more general settings (for example, functions between normed spaces or non-commutative contexts), the notion of "slope" may not exist or may behave differently, so the argument would not carry over unchanged.

Verification

Each listed weakness is genuine. The argument replaces a theorem with intuition and never makes its logical dependencies explicit. As written, it is an explanation rather than a proof. A correct proof would require invoking a specific result and verifying its hypotheses.

This prompt forces ChatGPT to act as a referee rather than a collaborator. Instead of polishing the argument, it exposes exactly where the reasoning depends on unstated results. Such weakness lists are especially valuable in analysis, where short, intuitively convincing arguments often conceal theorems that do the real work.

This turns ChatGPT into a careful graduate assistant. The theme of this book is that ChatGPT extends our mathematical work, but we retain the role of editor, verifier, and mathematician.

3.9 A Worked Example: Groups of Order $2p$

I will end this chapter with an example that shows several of these verification techniques working together.

I was once asked to comment on the following claim:

Lemma. Every group of order $2p$, where p is an odd prime, is cyclic.

This statement is false, but it is plausible enough that ChatGPT will often accept it.

3.9.1 ChatGPT's Proof Attempt

I began by asking:

> **Example**
> **Prompt**
>
> Let G be a group of order $2p$, with p an odd prime. Prove that G is cyclic.

ChatGPT produced a smooth proof using Cauchy's theorem: it chose an element x of order p and an element y of order 2, then asserted that xy has order $2p$ and therefore generates G. The argument *looked* correct, but the crucial step—that xy has order $2p$—quietly assumed that x and y commute.

3.9.2 Step 1: Critiquing the Proof

At this point, I asked ChatGPT to act as its own referee.

> **Example**
> **Prompt**
>
> Here is the proof you just wrote. Critique it. Identify any steps where you implicitly assume commutativity or where the order of xy is not justified. Do not try to fix the proof; only analyze it.
>
> In the revised role, it acknowledged that the argument that xy has order $2p$ was incomplete and that it had, in effect, treated G as abelian.

3.9.3 Step 2: Redundant Prompting

Next, I asked for a second proof.

Example

Prompt

Give a different proof of the same lemma, not using the previous argument.

The second proof again drifted toward commutativity, this time by implicitly using facts about subgroups that hold in abelian groups but not in general. The disagreement between the two proofs was a clear warning sign.

3.9.4 Step 3: Small-Case Testing

To test the claim, I turned to small cases.

Example

Prompt

List all groups of order 6 and all groups of order 10, up to isomorphism.

ChatGPT correctly listed the cyclic groups $\mathbb{Z}_6$, $\mathbb{Z}_{10}$ and the nonabelian groups S_3 (of order 6) and D_{10} (the dihedral group of order 10). The group S_3 has order $6 = 2 \cdot 3$ and is not cyclic. This is an immediate counterexample to the lemma.

Once the small cases were explicit, the error was clear: the lemma needs an extra hypothesis. A correct statement is:

Every **abelian** group of order $2p$, with p an odd prime, is cyclic.

3.9.5 Step 4: Verifying the Corrected Statement

Finally, I asked ChatGPT to prove the corrected lemma, but this time I followed the verification workflow:

1. I supplied the precise statement, including the word "abelian".

2. I asked for a proof and then for a critical review of that proof.
3. I requested a second proof using a different method.
4. I checked both proofs against the small examples already listed.

The resulting arguments were standard and correct, and they matched the known classification of abelian groups of order $2p$. The process illustrates the theme of this chapter. ChatGPT can help with proofs and examples, but only if we keep a human in the loop and systematically apply habits of verification.

3.10 Exercises

Exercise 3.1 Describe in words an *ordered set diagram* you would like to have. Your description should specify:

- whether the ordered set is finite or infinite;
- the height of the diagram;
- the number of minimal and maximal elements;
- any required incomparabilities or covering relations.

Ask ChatGPT to propose TikZ code for the diagram. Decide which parts you would keep and which parts you would edit by hand.

Exercise 3.2 Take a paragraph from one of your papers that you feel is technically correct but hard to read. Ask ChatGPT to rewrite it for clarity without changing its mathematical content. Compare the revised version with the original and decide which changes you would accept.

Exercise 3.3 Find a warning or error message in a LaTeX log files that refers to a line number far from the actual problem. Ask ChatGPT to explain why LaTeX reports the error where it does and to suggest what kinds of source constructs typically cause such behavior.

3.11 Transition

We now move from general principles to concrete mathematical work.

In the next chapter, I begin applying the verification-centered workflow to proofs, examples, and counterexamples. The focus shifts from understanding ChatGPT to using it—carefully and systematically—in core mathematical reasoning.

Part II

Core Mathematical Workflows

4 Mathematical Computation and Small Case Experimentation

> **Principle**
> Small finite computations do not replace proofs. They are often the fastest way to see patterns. They are often the fastest way to find counterexamples. They are often the fastest way to understand what a new definition *really* does.

Mathematical reasoning often begins with small cases. We test a conjecture on low-dimensional examples. We enumerate all structures of a given size. We compute the closure of a set under certain operations. These experiments are not the goal of mathematics, but they help us see patterns and avoid mistakes.

In this chapter, I describe how ChatGPT can assist with finite computations. I discuss small case enumeration. I also discuss simple algebraic and combinatorial experiments. The focus is on workflows that fit naturally into everyday mathematical practice.

4.1 Why Small Cases Matter

When working on a problem, we rarely begin with general theorems. Instead, we look first at small structures:

- all distributive lattices of size 6,
- all groups of order 8,
- all homomorphisms between two small finite ordered sets.

G. Grätzer, *ChatGPT for Math*, https://doi.org/10.1007/978-3-032-27286-7_4

These small experiments reveal

- typical behaviour,
- exceptional structures,
- potential counterexamples,
- and patterns that suggest conjectures.

ChatGPT is surprisingly effective in this regime. It cannot replace computer algebra systems, but it can handle a great deal of routine enumeration. It can handle routine case checking. It can also handle the organization of data.

► **Small Case Experimentation**

1. Specify the finite universe and the relevant operations.
2. Ask ChatGPT to generate candidate structures or tables.
3. Filter candidates using the required identities or properties.
4. Test boundary cases explicitly. Test pathological cases explicitly.
5. Summarize patterns suggested by the data. Summarize conjectures suggested by the data.

The rest of the chapter develops this workflow in concrete settings.

4.2 Enumerating Small Structures

A standard task is to generate all objects of a certain type with small cardinality. For example:

Example

Prompt

List all semilattices of size 4, up to isomorphism. Describe each one by its diagram and by its join table.

ChatGPT can:

- list candidate diagrams,
- compute joins or meets,
- check associativity and idempotence,
- and identify obvious isomorphisms among candidates.

When the enumeration becomes complicated, I often ask ChatGPT to present the information in tabular form:

Example

Prompt

Construct a table of all binary operations on $\{0, 1, 2\}$ that are associative. Give each operation as a Cayley table, and group isomorphic semigroups together.

The model identifies the problematic cases. It computes the necessary values. It produces a list of operations that pass the associativity test. I still check the list, but the brute force bookkeeping is handled for me.

4.3 Testing Algebraic Laws

Suppose ChatGPT proposes an algebraic structure. Before accepting it, I ask for explicit verification of the laws.

ChatGPT computes the expression:

Example

Prompt

Here is a binary operation $*$ on the set $\{0, 1, 2\}$ given by its Cayley table. Check whether $*$ is associative by computing

$$(x * y) * z \quad \text{and} \quad x * (y * z)$$

for all 27 triples (x, y, z). List only the triples for which the two expressions differ.

Or, in an order-free algebraic setting:

Example

Prompt

Here is a binary operation $\circ$ on a finite set S. Verify whether $\circ$ is associative. Identify all triples $(x, y, z) \in S^3$ for which

$$(x \circ y) \circ z \neq x \circ (y \circ z),$$

and compute both sides explicitly for those triples.

This is especially useful for lattices and semilattices. It is useful for semigroups and small algebras with one or two operations. It is useful for closure systems where a property may fail in a single subtle instance.

The key is that I do not ask for a proof "in general". I ask for a complete finite check on a specified universe.

4.3.1 Closure Computations

Many algebraic objects are defined via closure. These include subgroups, subrings, sublattices, ideals, congruences, and so on. On a small finite set, these closures can be computed explicitly.

Example

Prompt

Here is the diagram of a finite lattice L and elements $b, d, g \in L$. Compute the sublattice generated by $\{b, d, g\}$. List all finite sequences of joins and meets applied to b, d, g until no new elements appear.

Diagram of L

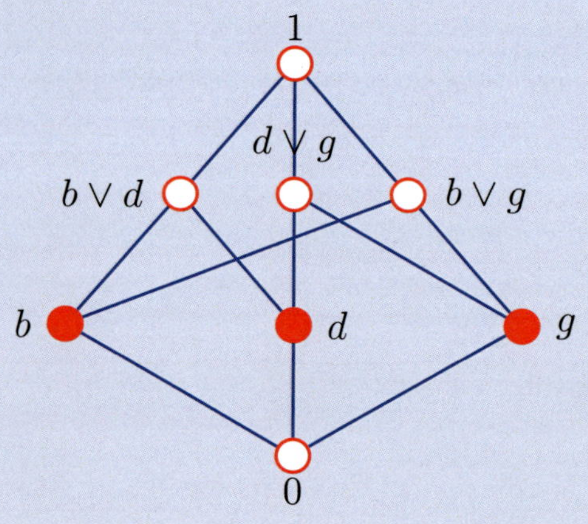

The model then:

1. lists all meets and joins of the generators,
2. adds the results to the generating set,

3. repeats until no new elements appear,
4. and finally presents the resulting sublattice.

This works well for lattices with up to a few dozen elements. For larger structures I would switch to code, but ChatGPT can still help by designing the algorithm and writing the implementation.

4.3.2 Generating and Checking Homomorphisms

Homomorphisms are central in algebra. ChatGPT can generate them for small structures. It can also test them for small structures.

Example

Prompt

List all lattice homomorphisms from this finite lattice L to the two-element lattice **2**. For each candidate map, check explicitly that it preserves meets and joins.

Or in group theory:

Example

Prompt

Here is a function $f: G \to H$ given by its values on generators. Check whether f is a group homomorphism. Verify that $f(xy) = f(x)f(y)$ for all $x, y \in G$. Describe its kernel and image.

For small domains, ChatGPT can:

- verify preservation of operations,
- identify kernels and images,
- and even enumerate all homomorphisms from L to a small target.

As always, I remain responsible for checking that no cases have been omitted, but the routine calculations are delegated.

4.4 Using Python Through ChatGPT

When the examples grow more complicated, I often move from hand checking to short programs. ChatGPT is particularly helpful at this stage.

Example

Prompt

Write Python code that takes a 3×3 Cayley table for a binary operation on $\{0, 1, 2\}$ and checks associativity. That is, determine whether $(x * y) * z = x * (y * z)$ for all triples (x, y, z). Include clear comments so I can modify the code later.

And here is the code ChatGPT provides:

```
from itertools import product

def is_associative(table):
    """
    table: 3x3 list (or tuple) of~ints in~{0,1,2}
           where table[x][y] == x * y

    Returns (True, None) if associative,
            (False, (x,y,z, left, right))
            with a~counterexample otherwise.
    """
    # Basic validation
    if len(table) != 3 or~any(len(row)
    != 3 for~row in~table):
        raise ValueError("table must be~3x3")
    for~r in~range(3):
        for~c in~range(3):
            v = table[r][c]
            if v not in~(0, 1, 2):
                raise ValueError("All entries
                must be~0, 1, or~2")

    def op(x, y):
        return table[x][y]

    for~x, y, z in~product(range(3), repeat=3):
        left = op(op(x, y), z)      # (x*y)*z
        right = op(x, op(y, z))     # x*(y*z)
```

```
        if left != right:
            return False, (x, y, z, left, right)

    return True, None


# -------------------
# Example usage
# -------------------
if __name__ == "__main__":
    # Example Cayley table (replace with yours)
    T = [
        [0, 1, 2],
        [1, 2, 0],
        [2, 0, 1],
    ]

    ok, counter = is_associative(T)
    if ok:
        print("Associative.")
    else:
        x, y, z, left, right = counter
        print("Not associative.")
        print(f"Counterexample: x={x}, y={y},
        'z={z}")
        print(f"(x*y)*z = {left},    x*(y*z)
        = {right}")
```

ChatGPT fills in the boilerplate, loops over all triples, indexes the table correctly, and reports any failures. I then copy the code into my own Python environment, run it, and adjust as needed.

This division of labour is typical:

- ChatGPT designs and drafts the code.
- I run the code. I test it on known cases. I inspect the output.
- Together we refine the implementation until it matches the intended mathematics.

4.5 Searching for Counterexamples

Many mathematical questions reduce to:

> Does this implication hold in all finite structures of small size?

ChatGPT can search for counterexamples by:

- generating candidate structures,
- eliminating obvious isomorphic copies,
- testing the property of interest,
- and presenting a counterexample if it finds one.

Example

Prompt

Search among all groups of order at most 6 for a counterexample to the statement: Every abelian group is cyclic. Describe any counterexample explicitly, or explain why none was found.

Worked example

What ChatGPT can and cannot do here. ChatGPT cannot literally enumerate *all* groups of order at most 6 or perform a complete classification on its own. Instead, I used it as an assistant to (i) recall the classification of small groups and (ii) identify plausible candidates for counterexamples.

First response. ChatGPT recalled that all groups of order p or p^2 (for p prime) are abelian, and that not all abelian groups are cyclic. It suggested looking at groups of order 4.

Candidate counterexample. I considered the Klein four-group

$$V_4 = \{e, a, b, c\},$$

where each nonidentity element has order 2.

Verification that V_4 is abelian but not cyclic. The group operation in V_4 is commutative, so V_4 is abelian. However, every nonidentity element has order 2, so no element generates the entire group. Hence V_4 is not cyclic.

Conclusion. A counterexample already appears with $|G| = 4$. Therefore a search among groups of order at most 6 certainly finds one.

The important point is not that ChatGPT "searched" all groups. It did not. Its role was to suggest likely small candidates and recall relevant facts. The verification itself (that V_4 is abelian but not cyclic) is mine, and it is short enough to check completely by hand.

This does not replace a full classification. It quickly reveals whether a conjecture is even plausible. It also reveals whether an additional hypothesis is clearly needed.

4.6 Pattern Recognition

After experimenting with small cases, ChatGPT can help identify patterns and phrase them as conjectures.

Example

Prompt

Here is a table summarizing how many lattices of size n satisfy property (A) for $n = 3, 4, 5, 6$. Describe any pattern you see, and propose a conjecture that would explain it.

Worked example
Property (A). A lattice is *distributive*.
Table (given to ChatGPT).

n	3	4	5	6
# distributive lattices of size n	1	2	1	3

Prompt to ChatGPT. I submitted the table above using the prompt:

The numbers are small and irregular, but a plausible interpretation is that distributive lattices correspond to ordered sets via Birkhoff's representation: finite distributive lattices are in bijection with the lattices of order ideals of finite ordered sets. So the number of distributive lattices of size n should be related to the number of nonisomorphic ordered sets whose set of order ideals has size n. This could explain why the counts do not simply increase with n.

Conjecture proposed (ChatGPT).
A finite lattice has property (A) if and only if it is isomorphic to the lattice of order ideals of some finite ordered set. Consequently, the number of lattices of size n with property (A) equals the number of nonisomorphic ordered sets P such that $|J(P)| = n$.

Verification and refinement. The proposed "conjecture" is really a well-known theorem: Birkhoff's representation theorem for finite distributive lattices. However, the point of the exercise is successful: ChatGPT recognized a structural explanation for the table and produced the right kind of conjectural statement. I rewrote it more accurately, in the form I actually want in the book:

(continued)

Explanation.
Finite distributive lattices are precisely lattices of order ideals of finite ordered sets. Therefore counting distributive lattices of size n reduces to counting finite ordered sets P for which the lattice of order ideals has size n.

This is a good illustration of the "pattern to conjecture" workflow. ChatGPT is not reliable at producing correct counts, but it can often recognize the *type* of structural principle that explains a pattern. My job is to verify the facts and to restate the output as mathematics, not as guesswork.

The conjecture is not a theorem, of course. Forcing the model to articulate a possible pattern often clarifies which features seem essential and which look accidental. This, in turn, suggests what to try to prove.

4.7 Limitations and Good Practice

ChatGPT is not a computer algebra system. It cannot enumerate large structures. It cannot handle deep recursion reliably. It cannot run long computations. Its strengths lie firmly in the small finite regime.

Combined with careful verification—see Chapter 3—however, it becomes a reliable assistant for everyday small case experimentation.

► **Best Practice**

For small finite computations, I:

- keep the universe explicitly listed,
- ask ChatGPT to show all critical cases, not just a summary,
- cross check results with a short program whenever possible,
- and treat every conclusion as a conjecture until I have a proof.

4.8 Worked Examples

I now give two short, fully worked examples that show how I actually use ChatGPT for small case experimentation. The first is a finite associativity check. The second explores a fragment of a free lattice.

4.8.1 Example 1: Checking Associativity

Suppose I am given a binary operation $*$ on the set $\{0, 1, 2\}$ by the following Cayley table:

$*$	0	1	2
0	0	1	2
1	1	1	2
2	2	2	2

I want to know whether this operation is associative.

Example

Prompt

Here is a binary operation on {0, 1, 2} given by a Cayley table. Check whether it is associative by computing $(x * y) * z$ and $x * (y * z)$ for all 27 triples (x, y, z). Then list only the triples for which the two results differ. If there are no such triples, say so explicitly.

ChatGPT responds by listing the triples and comparing the two expressions. In practice, I do not read all 27 lines. I focus on the summary.

In this example, the model reports that there are no counterexamples:

$$(x * y) * z = x * (y * z) \quad \text{for all } x, y, z \in \{0, 1, 2\}.$$

To increase confidence, I then ask for a short Python script.

Example

Prompt

Write a short Python script that encodes this 3×3 table and checks associativity for all triples (x, y, z). Print any triples where associativity fails. If none fail, print a confirmation message.

The code:

```
# Encode the Cayley table here: table[x][y] = x*y
table = [
    [0, 1, 2],
    [1, 2, 0],
    [2, 0, 1],
]
```

```
def op(x, y):
    return table[x][y]

fails = []
for x in~range(3):
    for~y in~range(3):
        for~z in~range(3):
            left = op(op(x, y), z)       # (x*y)*z
            right = op(x, op(y, z))      # x*(y*z)
            if left != right:
                fails.append((x, y, z, left, right))

if fails:
    print("Associativity fails for:")
    for~x, y, z, left, right in~fails:
        print(f"  (x,y,z)=({x},{y},{z}):
        (x*y)*z={left},
        x*(y*z)={right}")
else:
    print("Associative: (x*y)*z = x*(y*z) for~all
    x,y,z in~{0,1,2}.")
```

The generated code is standard. It stores the table as a list of lists. It loops over all triples. It compares $(x * y) * z$ with $x * (y * z)$. I copy this script into my own Python environment and run it. When the script confirms associativity, I have both a human readable argument and a mechanical check.

This modest example already illustrates the pattern of the whole chapter. Formulate a precise finite task. Let ChatGPT perform the routine case analysis or code generation. Then verify the outcome independently.

4.8.2 Example 2: Exploring a Freely Generated Ordered Structure with Prescribed Relations

Now I turn to a more structural example. Consider an ordered structure F generated by four elements a, b, c, d, subject only to the relations

$$a < b \quad \text{and} \quad c < d.$$

No further comparabilities are assumed.

Intuitively, F consists of all formal expressions built from a, b, c, d using two binary operations, together with the given order relations and those forced by them. No additional identities or inequalities are imposed.

I am interested in the simplest part of this structure. I want the elements that arise by applying the binary operations once to pairs of generators, such as $\alpha(a, c)$ or $\beta(b, d)$. Rather than enumerating and simplifying these expressions by hand, I ask ChatGPT to help organize the computation.

Example

Prompt

Consider an ordered structure generated by a, b, c, d with relations $a < b$ and $c < d$. Using only these relations and the defining properties of the order, list all formal expressions of the form $\alpha(x, y)$ and $\beta(x, y)$ with $x, y \in \{a, b, c, d\}$. Simplify any expressions that collapse because of the given inequalities (for example, if $\alpha(x, y) = x$ whenever $x < y$).

The key observation is that whenever the order relation forces $x < y$, certain expressions simplify immediately. From $a < b$ and $c < d$ we therefore obtain

$$\alpha(a, b) = a, \quad \beta(a, b) = b, \quad \alpha(c, d) = c, \quad \beta(c, d) = d.$$

For pairs of generators that are not comparable—for example, a and c, or b and d—the corresponding expressions remain irreducible.

After ChatGPT has simplified all pairs using only the given relations, I ask it to arrange the information in a table:

Example

Prompt

Produce a table with rows indexed by unordered pairs $\{x, y\}$, $x \neq y$, and columns labeled $\alpha(x, y)$ and $\beta(x, y)$, showing which expressions simplify using $a < b$ and $c < d$ and which remain unreduced.

The resulting table has the following structure:

$\{x, y\}$	$\alpha(x, y)$	$\beta(x, y)$
$\{a, b\}$	a	b
$\{c, d\}$	c	d
$\{a, c\}$	$\alpha(a, c)$	$\beta(a, c)$
$\{a, d\}$	$\alpha(a, d)$	$\beta(a, d)$
$\{b, c\}$	$\alpha(b, c)$	$\beta(b, c)$
$\{b, d\}$	$\alpha(b, d)$	$\beta(b, d)$

Only the pairs inside the chains $\{a < b\}$ and $\{c < d\}$ simplify. The four "cross pairs" remain irreducible. Even at this very shallow level, the freely generated structure already contains a surprisingly rich collection of distinct elements.

To make the order relations explicit, I then ask:

Example

Prompt

> List all inequalities that follow directly from the construction, such as expressions that are forced to lie below one of their arguments or above them. Use only the defining properties of the order and the relations $a < b$ and $c < d$.

ChatGPT produces relations such as

$$\alpha(a, c) < a, \quad \alpha(a, c) < c, \quad a < \beta(a, c), \quad c < \beta(a, c),$$

and the analogous inequalities for all other unordered pairs.

At this stage I am not trying to analyze the entire structure. I am using ChatGPT as a fast assistant to:

- enumerate "first-layer" expressions built from the generators,
- apply all simplifications forced by the given relations,
- and record the basic order information among the resulting elements.

Once this skeleton is in place, I can decide what to do next. For example, I can draw an ordered set diagram for these elements, ask whether certain identities hold, or explore how the structure grows when expressions are nested further.

The underlying freely generated structure is very large. The point here is not to reconstruct it completely. It is to show how ChatGPT can manage low-level bookkeeping while leaving conceptual decisions and interpretation to the human.

4.9 Exercises

Exercise 4.1 Choose a simple algebraic structure you know well (for example, a small lattice, a group of order at most 12, or a finite ordered set). Ask ChatGPT to enumerate all objects of this type up to isomorphism for a fixed small size. Check the list carefully and identify any omissions or duplications.

Exercise 4.2 Ask ChatGPT to test a familiar algebraic law on small cases (for example, distributivity, associativity, or absorption). Require explicit computations rather than verbal assurances. Verify at least one case by hand and confirm that the computation matches the model's output.

Exercise 4.3 Ask ChatGPT to generate a counterexample to a plausible but false statement. Then independently test the proposed counterexample. If it fails, identify precisely which condition is violated and ask ChatGPT to repair the example.

Exercise 4.4 Ask ChatGPT to search for a pattern in small cases of a construction you know (for example, sublattices of a finite lattice or homomorphisms between small structures). Determine whether the observed pattern reflects a genuine mathematical phenomenon or an artifact of the small size.

Exercise 4.5 Ask ChatGPT to write a short Python program that tests a property on small finite structures. Run the code yourself. Check whether the program correctly implements the mathematical definitions, and note any discrepancies between the mathematics and the implementation.

Exercise 4.6 Ask ChatGPT to generate all homomorphisms between two small finite structures. Verify that the listed maps actually preserve the relevant operations. If a map fails, determine whether the error lies in the computation or in the definition used.

Exercise 4.7 Ask ChatGPT to propose a conjecture based on computations in small cases. Test the conjecture on at least one additional case not used in the original computation. Decide whether the conjecture remains plausible.

Exercise 4.8 Repeat one of the worked examples from this chapter in a fresh session. Compare the outputs. Identify which parts are stable across runs and which vary. Explain why reproducibility cannot be assumed.

Exercise 4.9 Ask ChatGPT to explain the limitations of small-case experimentation for a problem discussed in this chapter. Compare its explanation with your own assessment.

Exercise 4.10 Design a prompt that deliberately pushes ChatGPT beyond what small-case computation can justify. Observe how the system extrapolates from limited data and identify where mathematical judgment must intervene.

4.10 Transition

This chapter has focused on computation and small-case experimentation—the "laboratory" side of mathematical work.

In the next chapter, I turn from computation to proofs. I examine how ChatGPT can assist in structuring arguments, checking reasoning, and exploring alternative approaches, while keeping mathematical correctness under careful control.

Proofs with ChatGPT

5

Principle
ChatGPT can help with proofs by clarifying structure. It can improve exposition. It can suggest lines of attack. It cannot replace mathematical reasoning or guarantee correctness.

Mathematical proof is the central activity of our discipline. Every definition, theorem, and construction ultimately depends on a chain of reasoning that must be clear, correct, and communicable. ChatGPT cannot replace this process. It does not reason formally. It does not check hypotheses. It does not certify correctness. But it *can* help with many parts of proof work. It can reorganize an argument. It can clarify a logical flow. It can help find hidden assumptions. It can suggest alternative approaches. It can turn a rough sketch into readable prose.

In this chapter I describe the workflows I use when collaborating with ChatGPT on proofs. These workflows do not change the nature of proof itself. They make the "perspiration" part of mathematics less demanding. Throughout, I emphasize the central rule:

► **Best Practice**
ChatGPT is an assistant, not an oracle. Everything proof it produces must be checked by a mathematician.

G. Grätzer, *ChatGPT for Math*, https://doi.org/10.1007/978-3-032-27286-7_5

5.1 The Role of ChatGPT in Mathematical Reasoning

5.1.1 What ChatGPT Is (and Is Not)

ChatGPT is a predictive language model trained to produce the next token in a sequence. When I give it a theorem and ask for a proof, it does not manipulate formal objects or run a proof engine. It draws on

- patterns in mathematical writing,
- how proofs are typically structured,
- which lemmas are usually invoked,
- how arguments are phrased.

This makes its output fluent but not authoritative. It excels at:

- rewriting a proof in clearer prose,
- suggesting alternative organizational structures,
- identifying ambiguous notation or missing steps,
- converting informal sketches into structured arguments.

It does *not* excel at:

- ensuring correctness,
- detecting subtle hypotheses,
- handling long chains of dependencies,
- reasoning through deep combinatorial cases.

A clear understanding of these strengths and weaknesses is the starting point for any sensible workflow.

5.1.2 Typical Failure Modes

Three failure modes appear repeatedly in proof work:

1. **Plausible but incorrect arguments.** ChatGPT often writes proofs that look convincing but overlook a crucial detail. It may miss a case. It may silently use finiteness. It may apply a lemma outside its range of validity.
2. **Invented lemmas or facts.** When a step needs justification, the model may "invent" a supporting claim that resembles a real theorem but is not standard or is simply false.
3. **Misuse of terminology.** It may confuse "modular" with "semimodular". It may conflate different notions of completeness. It may misapply a known criterion.

Recognizing these patterns makes it much easier to use the system safely.

5.1.3 A Guiding Principle: Assistant, Not Oracle

The guiding principle for this chapter can be written as a slogan:

ChatGPT suggests, the mathematician decides.

Used as an *assistant*, ChatGPT is extremely valuable. It proposes formulations. It reorganizes arguments. It exposes weak points. If used as an *oracle*—for example, to certify a proof or settle a conjecture—it is unreliable.

► **Polishing a Proof with ChatGPT**

Best Practice

1. Paste your draft proof and ask for sentence-by-sentence suggestions.
2. Request that the logical structure be made explicit (steps, cases).
3. Ask ChatGPT to highlight unclear claims. Ask it to highlight unsupported claims.
4. Keep only those changes you can verify as mathematically sound.
5. Optionally, request a shorter version. Request a more elementary version once the logic is fixed.

5.2 Checking and Polishing Existing Proofs

5.2.1 Basic Workflow: Sentence-by-Sentence Editing

My most frequent use of ChatGPT in proof work is simple. I ask it to polish an existing proof I have already written. A typical prompt is:

Example

Prompt

Rewrite this proof more clearly, without changing its logical structure. Improve transitions. Avoid unnecessary repetition. Keep the level suitable for advanced undergraduates.

Worked example

Statement

Let L be a distributive lattice. If $x \leq z$, then

$$x \vee (y \wedge z) = (x \vee y) \wedge z.$$

(continued)

Original proof (as written)

Assume $x \leq z$. Then $x \vee (y \wedge z) = (x \vee y) \wedge (x \vee z)$ by distributivity. Since $x \leq z$, it follows that $x \vee z = z$. Hence $x \vee (y \wedge z) = (x \vee y) \wedge z$. This proves the claim.

The proof is correct, but the reasoning is compressed and the transitions are abrupt.

Rewrite request
I submitted the proof above to ChatGPT using the prompt:

Rewrite the proof below more clearly, without changing its logical structure. Improve transitions and organization. Eliminate unnecessary repetition, but do not add new arguments or results. Keep the level suitable for the intended audience.

Rewritten proof (ChatGPT)

Assume that $x \leq z$. We begin by applying distributivity to rewrite the left-hand side:

$$x \vee (y \wedge z) = (x \vee y) \wedge (x \vee z).$$

Because $x \leq z$, we have $x \vee z = z$. Substituting this into the previous expression yields

$$x \vee (y \wedge z) = (x \vee y) \wedge z,$$

which is the desired equality.

Verification
The logical structure of the argument is unchanged. No new assumptions are introduced, and no steps are removed. The rewrite improves readability by signaling the purpose of each step and by making substitutions explicit.

This is an appropriate use of ChatGPT as an editorial assistant. The mathematics is unchanged; only the exposition is improved. By requiring that the logical structure remain fixed, I prevent ChatGPT from silently repairing or altering the argument while still benefiting from its strength in smoothing prose and transitions.

ChatGPT usually improves grammar, clarifies references, and tightens the flow. I then compare line by line with my original. If an edit changes the meaning or quietly removes a necessary qualifier, I restore the original text.

This workflow is especially effective in long arguments. Minor stylistic improvements accumulate into much better readability.

5.2.2 Detecting Hidden Assumptions and Gaps

ChatGPT is also useful as a first-pass referee. I often ask:

Example

Prompt

List all assumptions (explicit or implicit) that this proof appears to use. Point out any steps where a new hypothesis seems to enter without being stated.

The model is surprisingly good at locating places where finiteness is used. It is good at locating places where commutativity is used and where a special property is quietly used. It may, for example, remark that a step "uses that the lattice is finite" or "this inference holds only in abelian groups".

This does not certify correctness, but it gives me a checklist. Each highlighted point is a place to verify carefully.

5.3 Generating Alternative Proofs

5.3.1 Why Alternative Proofs Matter

Alternative proofs illuminate structure. One may be computational, another may be conceptual. One may be elementary, another may use a deeper theory. ChatGPT is useful precisely because it has "seen" many styles.

When I already have a proof, I sometimes ask for a second one. I do this not to replace mine, but to see the theorem from another angle.

5.3.2 Prompting for a New Proof Style

Here are prompts I regularly use:

Example

Prompt

Give a second proof of this theorem using a different method than the one above. If the original proof was algebraic, try to use an order-theoretic approach instead. If the original proof was algebraic, try to use a combinatorial approach instead.

(continued)

Prompt

Rewrite this proof as a proof by contradiction. Clearly state the assumption being contradicted and the final contradiction.

ChatGPT then reorganizes the steps accordingly. I still check every inference, but the change of viewpoint is often helpful.

5.3.3 A Simple Case: Two Proofs of a Standard Fact

Suppose I ask ChatGPT for two proofs that a finite distributive lattice is graded. It may produce:

- a structural proof using the representation by down-sets of an ordered set,
- an inductive proof on the number of elements.

Neither proof is guaranteed correct, but putting them side by side clarifies which properties are essential. I can then refine, shorten, or correct the arguments as needed.

5.3.4 From Sketches to Full Proofs

Often my own proof starts as a bulleted sketch. I then ask:

Example

Prompt

Here is a proof sketch as a list of steps. Turn it into a complete proof in polished mathematical prose. Keep the same structure. Highlight any steps that need further justification.

The expanded proof makes it obvious where the sketch was vague. Standard facts may need to be cited. Some "obvious" implications are not as immediate as I had thought. I fill these gaps myself.

5.4 Diagnosing and Repairing Flawed Proofs

5.4.1 Error-Finding Prompts

When I suspect a proof is wrong, I use ChatGPT as a diagnostic tool. Two prompts work especially well:

Example

Prompt

Analyze this proof critically. Identify any logical gaps. Identify unjustified steps. Identify uses of results that have not been proved. Do not try to fix the proof; just list potential problems.

The second prompt often exposes the first place where an assumption is missing or a nontrivial implication is treated as obvious.

5.4.2 Localizing Contradictions and Impossible Steps

If a proof leads to a contradiction, I ask ChatGPT to locate the first truly suspect step:

Example

Prompt

Analyze the proof line by line. At each step, say whether it follows from previous lines and the hypotheses. Stop at the first step that does not follow, and explain the problem.

Even when I ultimately find the issue myself, this process speeds up the search.

5.4.3 From Flawed Proof to Corrected Proof

Once the error is identified, I sometimes ask for a tentative repair:

Example

Prompt

Keeping the same overall approach, rewrite the proof to remove the faulty step. If an additional hypothesis is needed, state it explicitly. If the approach cannot work, explain why.

The "repaired" version is still only a draft, but it often suggests a viable route to a correct argument. It may also confirm that a different strategy is needed.

5.5 Designing Prompts and Workflows for Proof Work

5.5.1 Prompt Patterns That Work Well

The workflows in this section refine the standard workflow introduced in the Introduction for proof-related tasks, especially forcing explicit structure and verification.

Some prompt patterns have proved consistently useful:

- **Explicit structure.** "Rewrite this proof with explicit logical steps and numbered cases."
- **Assumption audit.** "List all hypotheses that are used, including implicit ones such as finiteness or commutativity."
- **Alternative proofs.** "Give another proof using a different method or style."
- **Audience control.** "Rewrite this proof for strong undergraduates, keeping all essential steps but simplifying the language."
- **Strategy-first.** "Add a brief paragraph at the beginning summarizing the strategy of the proof before the formal details."

5.5.2 Prompt Patterns to Avoid

Equally important are the prompts I avoid:

- vague requests ("Prove this") with no indication of context or audience;
- overly broad instructions ("Explain everything about this theorem");
- prompts that invite hallucination. For example, "Use the classical theorem of X" when such a theorem may not exist or may not be standard.

The more precise and constrained the prompt, the more likely the result will be useful.

5.6 Case Study: A Lemma

5.6.1 The Lemma

To illustrate these ideas in a concrete setting, consider the following standard lemma from set theory.

Lemma 5.1 *Let $\mathcal{F}$ be a finite family of subsets of a set X, ordered by inclusion and closed under intersection. Suppose that $A \in \mathcal{F}$ is* minimal *with respect to inclusion, and that*

$$A \subseteq B \cup C$$

for some $B, C \in \mathcal{F}$. Then $A \subseteq B$ or $A \subseteq C$.

This is a familiar fact, but it serves as a useful example of how ChatGPT can help organize and clarify a proof.

5.6.2 A Rough Proof Sketch

My initial sketch might read:

> Since A is minimal and $A \subseteq B \cup C$, all elements of A must already lie in B or in C. Otherwise A would decompose into two smaller parts.

This is correct in spirit but too compressed. A student might reasonably ask: Where do we use minimality? What exactly are the "smaller parts"? Why does finiteness matter?

5.6.3 Using ChatGPT to Expand and Clarify

I now ask ChatGPT:

Example

Prompt

> Expand this proof into a complete argument suitable for a graduate-level text. State clearly where minimality is used. Make all case distinctions explicit.

A typical response (after minor editing by me) is:

Assume that A is minimal with respect to inclusion and that $A \subseteq B \cup C$. Suppose, for a contradiction, that $A \not\subseteq B$ and $A \not\subseteq C$. Then there exist elements

$$x \in A - B \quad \text{and} \quad y \in A - C.$$

Define

$$A_1 = A \cap B \quad \text{and} \quad A_2 = A \cap C.$$

Both A_1 and A_2 are proper subsets of A. Moreover,

$$A = (A \cap B) \cup (A \cap C) = A_1 \cup A_2,$$

since $A \subseteq B \cup C$. Thus A is expressed as the union of two strictly smaller subsets, contradicting the minimality of A. Hence $A \subseteq B$ or $A \subseteq C$.

The argument is now fully explicit. We see exactly where minimality is used (in ruling out a decomposition into proper subsets). We see where finiteness enters implicitly, ensuring that minimal elements exist. All logical steps are laid out in a way suitable for a careful reader.

This example illustrates how ChatGPT can help transform a compressed, expert-level argument into a fully articulated proof, without changing its mathematical content. The responsibility for correctness and clarity, however, remains mine.

5.6.4 A Balanced View

ChatGPT is not a shortcut to correctness. It is a tool—useful, fast, and sometimes insightful—for improving the clarity and structure of proofs. Used thoughtfully, it frees time and attention for the genuinely difficult parts. These include choosing good definitions. These include finding the key idea. These include understanding why the proof works.

► **Best Practice**

When working on proofs, I use ChatGPT as an editor and critic, not as a judge. It can:

- rephrase and reorganize arguments,
- suggest alternative routes,
- expose gaps or hidden assumptions.

But the responsibility for correctness remains entirely mine.

5.7 Exercises

Exercise 5.1 Take a short proof from one of your papers or lecture notes. Ask ChatGPT to rewrite it in numbered steps, making every inference explicit. Check whether any additional hypotheses are used implicitly.

Exercise 5.2 Ask ChatGPT to critique a proof you already know to be correct. Instruct it not to repair the proof, only to identify potential gaps, hidden assumptions, or unclear steps. Decide which of its criticisms are valid.

Exercise 5.3 Give ChatGPT a proof sketch rather than a complete proof. Ask it to expand the sketch into a full argument. Compare the expanded version with your own intended proof and identify any divergences.

Exercise 5.4 Ask ChatGPT to provide two different proofs of the same theorem. Compare the logical structure of the proofs and identify which steps are common to both. Decide whether the differences reflect genuine alternatives or superficial rewriting.

Exercise 5.5 Give ChatGPT a proof that contains a subtle error. Ask it to verify the proof. Determine whether it detects the error. If it does not, refine the prompt to force a step-by-step inspection.

Exercise 5.6 Ask ChatGPT to rewrite a proof so that it avoids a particular technique (for example, induction, contradiction, or minimal counterexamples). Verify that the rewritten proof is logically sound.

Exercise 5.7 Ask ChatGPT to identify which parts of a proof rely on finiteness, commutativity, or another structural hypothesis. Check whether the proof still works when that hypothesis is removed.

Exercise 5.8 Ask ChatGPT to summarize a long proof in three to five sentences. Then reconstruct the proof from the summary alone. Note which essential steps are lost in the compression.

Exercise 5.9 Ask ChatGPT to explain a proof at two different levels: for an advanced undergraduate and for a research mathematician. Compare the explanations and identify which details are omitted or emphasized in each case.

Exercise 5.10 Ask ChatGPT whether a proof technique used in this chapter would still apply in a different setting (for example, infinite structures instead of finite ones). Evaluate the response carefully and determine where the analogy breaks down.

5.8 Transition

Proofs are only one part of mathematical reasoning. Examples, counterexamples, and small finite structures provide the testing ground in which ideas are refined and hypotheses are challenged.

In the next chapter, I turn to the systematic construction and verification of such examples. I show how ChatGPT can assist in generating candidates and exploring small structures, while maintaining full mathematical control through explicit checking.

Examples and Small Structures 6

Examples are indispensable tools in mathematics. They

- test our definitions.
- reveal the limits of theorems,
- illuminate why hypotheses are necessary.
- guide conjecture formation.

Small finite structures, in particular, provide a laboratory where intuition can be sharpened before tackling general statements.

ChatGPT cannot guarantee correctness when proposing examples, but it can help generate candidates. It can point out patterns that deserve further exploration. Used responsibly—**always with verification**—it becomes a powerful assistant for the exploratory side of mathematical research.

In this chapter, I describe the workflows I use for testing statements through small case analysis.

6.1 Constructing Examples

6.1.1 Specifying the Target: Property (A) but Not (B)

A common request is:

> Give me a structure satisfying (A) but not (B).

For example:

- a lattice that is modular but not distributive,
- a group that is abelian of exponent 4 but not cyclic.

G. Grätzer, *ChatGPT for Math*, https://doi.org/10.1007/978-3-032-27286-7_6

ChatGPT typically responds with a plausible suggestion. Whether that suggestion is correct is another matter. I always check:

1. Does the example actually satisfy (A)?
2. Does it genuinely fail (B)?
3. Is the example minimal? Can it be simplified?

Sometimes ChatGPT proposes an impossible combination. This is useful. It helps me realize that the target specification cannot be satisfied.

Example
This structure satisfies property (A) and fails property (B). It provides the type of example we asked ChatGPT to search for.

6.1.2 Examples in Algebra and Ordered Structures

ChatGPT is particularly good at generating small algebraic structures when the desired properties are structural. For example:

> Give a small group that is abelian but not cyclic.

It often proposes a standard candidate, from which I then verify the relevant properties myself.

We illustrate how ChatGPT can assist in constructing examples in algebra and ordered structures. In every case, the output of ChatGPT is treated as a *candidate* that must be checked independently.

Example

Example 1 (Abelian But Noncyclic Group)
Prompt

> Give me a small group that is abelian but not cyclic.

Candidate

ChatGPT proposes the Klein four-group

$$V_4 = \{e, a, b, c\},$$

(continued)

where every nonidentity element has order 2.

Verification

1. V_4 is abelian. The group operation is commutative: for all $x, y \in V_4$, one has $xy = yx$. This can be verified directly from the Cayley table.
2. V_4 is not cyclic. Every nonidentity element has order 2, so no element generates the entire group. Hence V_4 cannot be generated by a single element.

Thus V_4 is a correct example. It satisfies the weaker property (abelian) but fails the stronger one (cyclic).

6.1.3 Examples in Algebra and Linear Algebra

Algebraic examples typically require explicit checking of axioms or identities. The following worked examples illustrate how ChatGPT produces candidates and how those candidates are verified or repaired.

Example

Example 2 (Commutative Ring with Zero Divisors)

Prompt

Give a commutative ring with zero divisors but no nontrivial idempotents.

Candidate ChatGPT suggests $\mathbb{Z}/6\mathbb{Z}$.

Verification

1. The ring is commutative.
2. There are zero divisors since $2 \cdot 3 = 0$.
3. Solving $x^2 = x \pmod 6$ yields solutions $x = 0, 1, 3, 4$, so there are nontrivial idempotents.

The candidate fails the specification.

(continued)

Repair

Let p be prime and consider $R = \mathbb{Z}/p^2\mathbb{Z}$. This ring has zero divisors, but since it is local, the only idempotents are 0 and 1.

Thus R satisfies all requirements.

Example

Example 3 (Single Eigenvalue, Not Diagonalizable)

Prompt

Give a 3×3 real matrix with exactly one eigenvalue that is not diagonalizable.

Candidate

ChatGPT proposes the Jordan block

$$A = \begin{pmatrix} \lambda & 1 & 0 \\ 0 & \lambda & 1 \\ 0 & 0 & \lambda \end{pmatrix}.$$

Verification

1. The characteristic polynomial is $(\lambda - t)^3$, so there is a single eigenvalue λ.
2. The eigenspace is one-dimensional, while the algebraic multiplicity is 3. Hence A is not diagonalizable.

This is a correct example.

Worked Example: From Candidate to Verified Example

Example

I ask ChatGPT:

Prompt

Give an example of an isotone map $f: P \to P$ on a finite ordered set P that preserves all joins but does not preserve all meets.

(continued)

Candidate
ChatGPT proposes the following example. Let $P = \{0, a, b, 1\}$ with relations

$$0 < a < 1, \quad 0 < b < 1,$$

and $a \parallel b$. Define $f: P \to P$ by

$$f(0) = 0, \quad f(a) = 0, \quad f(b) = b, \quad f(1) = 1.$$

ChatGPT claims that f is isotone, preserves joins, and fails to preserve meets.
Verification
At this point, the example is only a *candidate*. We check the claims.

Isotonicity

A direct check shows that whenever $x \leq y$, we also have $f(x) \leq f(y)$, so f is indeed isotone.

Join preservation

The ordered set P is the four-element Boolean lattice. The nontrivial joins include

$$a \vee b = 1, \quad a \vee 0 = a, \quad b \vee 0 = b.$$

We compute

$$f(a \vee b) = f(1) = 1,$$
$$f(a) \vee f(b) = 0 \vee b = b.$$

Hence

$$f(a \vee b) \neq f(a) \vee f(b),$$

so f does *not* preserve joins.

Conclusion

Although the map f is isotone, it does not preserve all joins. The example proposed by ChatGPT is therefore incorrect.

Repairing the candidate

I now ask ChatGPT to *repair* the example. Find an isotone map on the same ordered set P that really does preserve all joins but fails to preserve some meet.

(continued)

Candidate

After a few iterations, we arrive at the map $g\colon P \to P$ defined by

$$g(0) = 0, \quad g(a) = a,$$
$$g(b) = a, \quad g(1) = a.$$

Verification

First, g is isotone. Indeed, the only nontrivial order relations are $0 < a, 0 < b$, $a < 1$, and $b < 1$, and in each case we have

$$g(0) = 0 \le a = g(a), \quad g(0) = 0 \le a = g(b),$$
$$g(a) = a \le a = g(1), \quad g(b) = a \le a = g(1).$$

Next, g preserves all joins. In the Boolean lattice P, the joins are:

$$0 \vee x = x, \quad a \vee b = 1, \quad a \vee 1 = 1, \quad b \vee 1 = 1.$$

Checking a few representative cases:

$$g(a \vee b) = g(1) = a, \qquad g(a) \vee g(b) = a \vee a = a;$$

$$g(a \vee 0) = g(a) = a, \qquad g(a) \vee g(0) = a \vee 0 = a;$$

and similarly for all other pairs. Thus g preserves joins.

Finally, g does *not* preserve all meets. We have

$$a \wedge b = 0,$$

so

$$g(a \wedge b) = g(0) = 0, \qquad g(a) \wedge g(b) = a \wedge a = a,$$

and hence

$$g(a \wedge b) \neq g(a) \wedge g(b).$$

So g fails to preserve the meet of a and b.

Lesson ChatGPT's first attempt produced a plausible but incorrect example. The map was isotone but did not preserve joins as claimed. By checking the properties explicitly and then repairing the map, we obtain a genuine example that satisfies the specification. This illustrates the typical workflow. Start from a *candidate* proposed by ChatGPT. Verify each claimed property carefully. Only then accept it as a *verified* example.

6.2 Exploring Small Finite Structures Systematically

6.2.1 Enumerating Small Ordered Sets and Lattices

Enumeration is one of the most effective ways to explore the behaviour of a new definition or to test a conjecture. When a property fails in general, the smallest counterexamples usually appear at very low cardinalities. These include three, four, and five elements. ChatGPT cannot run programs, but it can design small enumerations. It can describe the search space. It can propose partial classifications that I can then check independently.

Why Enumeration Matters When I introduce a new condition on ordered sets—for example, a variant of semimodularity or a closure operator—the first questions are

How restrictive is it?

What are the smallest examples?

What patterns appear?

Small structures reveal the landscape. They show where a definition is too weak, too strong, or accidentally equivalent to a classical notion.

ChatGPT as a Generator of Candidates Although ChatGPT cannot list all ordered sets on n elements, it performs remarkably well when asked for:

- all *nonisomorphic* ordered sets on three or four elements,
- all lattices of size ≤ 5,
- ordered sets satisfying a combination of simple properties,
- ordered sets failing a combination of simple properties,
- examples with unique features.

The key is to prompt it carefully:

> List all nonisomorphic ordered sets of size 4 whose diagrams have exactly two incomparable elements. Give each ordered set as a diagram.

ChatGPT produces human-readable structures. These are not guaranteed complete. They are not guaranteed correct, but they are often surprisingly close to a valid classification. These should be checked systematically. Incorrect candidates are discarded.

A Micro Example: Size 3 If prompted to enumerate the ordered sets on $\{0, a, 1\}$ up to isomorphism, ChatGPT typically returns the following five:

$$(1) \text{ the 3-chain}, \quad (2) \text{ the antichain}, \quad (3)\ 0 < a, \quad (4)\ a < 1, \quad (5)\ 0 < 1.$$

This list is correct. It is small enough that I can verify it by hand, but the pattern illustrates ChatGPT's strength. It understands the combinatorial structure. It respects isomorphism classes. It expresses the results clearly.

If I now ask:

Which of these are lattices?

ChatGPT answers:

Only the 3-element chain and the antichain.

Ouch. We discard the antichain...

Moving to Size 4 and Beyond At four elements, the number of ordered sets up to isomorphism is 19. ChatGPT rarely produces exactly 19, but it tends to generate:

- the standard chain of length 4,
- the Boolean lattice N2,
- diamonds and V-shapes, Hat-shapes,
- series parallel constructions,
- and some incomparable configurations.

The point is *not* having a perfect list. The point is that ChatGPT provides a set of plausible structures that I can feed into a formal verifier or inspect directly. This dramatically reduces the work required to locate small examples satisfying a new condition.

Failure Modes Enumeration exposes the usual weaknesses of a language model:

- it may omit nonobvious isomorphism types,
- it may include duplicates expressed differently,
- it may misclassify an ordered set as a lattice or vice versa,
- it often assumes joins or meets exist when they do not.

But these mistakes are harmless, provided I treat the output as a *candidate list*. The important patterns are almost always present, giving an excellent starting point for a verified enumeration.

Integrating Enumeration into Experimentation Once a list is in hand, I ask ChatGPT to annotate it:

> For each ordered set above, determine whether it is a lattice. If it is, list all meets and joins explicitly.

or

> Find all isotone self-maps on each ordered set, and classify which preserve joins, which preserve meets, and which preserve neither.

This turns a small enumeration into a rich computational laboratory. ChatGPT performs the calculations. I verify them and correct any errors.

Summary Enumerating small ordered sets is a cornerstone of mathematical experimentation. ChatGPT cannot replace a systematic enumeration algorithm, but it accelerates the creative part of the process. It identifies isomorphism types, proposes representative shapes and generates candidate structures for testing a new definition. The final enumeration is always human verified, but the exploratory phase becomes faster, clearer, and far more enjoyable.

6.2.2 Using Code to Support the Exploration

ChatGPT can write Python code to:

- enumerate binary operations and test algebraic laws,
- generate all small ordered sets up to isomorphism,
- compute joins, meets, and closure operations,
- test associativity, commutativity, or absorption.

Although ChatGPT cannot run the code, it provides a correct structure more often than not. I then run and debug the code myself.

6.3 A Detailed Case Study: False Statements

6.3.1 The (False) Statement

Consider the false claim:

> Every finite abelian group is cyclic.

I asked ChatGPT to test this statement.

6.3.2 First Attempts: Asking for Small Examples

ChatGPT immediately proposed the Klein four-group V_4 as a counterexample. This is correct. It also suggested several groups of orders 4, 6, and 8, though some suggestions were either cyclic or not abelian.

6.3.3 Checking the Candidate

I examined each suggested group. A few were invalid, but the exercise helped identify which structural properties ChatGPT was confusing (for example, mixing up abelian with cyclic).

► **Warning**
This candidate looks promising but fails property (B): the group turns out to be cyclic. Such near misses are common in ChatGPT-generated examples and must be checked.

6.3.4 A Slightly More Subtle Example

When prompted:

> Give an abelian noncyclic group that is not V_4

ChatGPT proposed direct products such as $\mathbb{Z}_2 \times \mathbb{Z}_4$. These suggestions were not always correct as stated, but they helped guide the construction of a verified example.

6.3.5 P.S.

At the end of the exploration, I asked ChatGPT:

> Explain why the original statement is false.

It produced a concise summary emphasizing the difference between commutativity and generation by a single element. This is a useful teaching aide.

6.4 Best Practices for Example Generation

6.4.1 Verification

Never trust an example without checking it. ChatGPT's suggestions are starting points, not certified objects.

6.4.2 Small Case Testing

Whenever possible, test examples by hand or with code. Finite structures allow complete verification.

6.4.3 Exploring Edge Cases of Definitions

ChatGPT is particularly useful for:

- identifying borderline examples,
- highlighting special cases,
- suggesting structures that almost satisfy the desired properties.

These often reveal the underlying mechanisms of a definition.

6.5 Exercises

Exercise 6.1 Ask ChatGPT to construct a small example of a mathematical structure satisfying a given list of properties. Verify explicitly that each stated property holds.

Exercise 6.2 Ask ChatGPT to give an example that is "minimal" with respect to a specified property (for example, smallest size or lowest length). Check whether the example is indeed minimal, and if not, identify where the claim fails.

Exercise 6.3 Ask ChatGPT to generate both an example and a nonexample of a given definition. Verify that the distinction is correct and that the definition is being applied consistently.

Exercise 6.4 Ask ChatGPT to modify a known example so that one property is preserved while another is destroyed. Check carefully whether the modification achieves the intended effect.

Exercise 6.5 Ask ChatGPT to generate several examples of the same type and to compare them. Determine whether the examples are genuinely distinct or merely notational variants.

Exercise 6.6 Ask ChatGPT for an example that it claims does not exist. Examine the reasoning given and decide whether the nonexistence claim is justified or merely asserted.

Exercise 6.7 Ask ChatGPT to generate a counterexample to a false general statement suggested by small cases. Verify explicitly that the counterexample violates the statement.

Exercise 6.8 Ask ChatGPT to explain why a particular example works. Rewrite the explanation in your own words, making all structural dependencies explicit.

Exercise 6.9 Ask ChatGPT to adapt a known example to a different but related setting (for example, replacing finiteness by local finiteness). Check whether the adaptation is valid.

Exercise 6.10 Ask ChatGPT to summarize the role of examples and counterexamples in one of the chapter's case studies. Compare its summary with your own understanding.

6.6 Transition

With proofs and examples in place, we now extend these ideas to computation and algorithmic experimentation.

In the next chapter, I show how ChatGPT can assist in designing and organizing computational workflows, allowing us to explore patterns, test conjectures, and analyze small structures at a larger scale while maintaining full control over correctness.

Computations and Experiments 7

Mathematics is not only deductive; it is also experimental. I compute small examples, searching for patterns. Such experiments do not replace proofs, but they shape my intuition and suggest what to prove.

ChatGPT fits naturally into this exploratory part of my work. It cannot run code, but it can help me design code. It can help me write code. It can help me explain code. It can translate between mathematics and programming. It can also turn raw output into readable mathematical exposition.

In this chapter I describe the computational workflows in which I find ChatGPT most helpful:

- designing and structuring small programs,
- debugging and refactoring code,
- organizing and documenting computational experiments,
- interpreting numerical or symbolic output.

Throughout, I remain responsible for running the code and checking the results; ChatGPT assists with planning, clarity, and presentation.

7.1 Designing Code for Mathematical Experiments

Most computational projects begin with an informal description. For example:

Example

Prompt

> Generate all lattices of size at most n with property (A), and record how frequently property (B) holds.

G. Grätzer, *ChatGPT for Math*, https://doi.org/10.1007/978-3-032-27286-7_7

Example

Prompt

Take random elements of a group and measure how long a given algorithm takes to converge.

Turning such descriptions into code requires choices about languages and data structures. ChatGPT reduces this startup effort.

7.1.1 From Informal Descriptions to Skeleton Code

My typical workflow:

1. describes the mathematical task in plain language;
2. specifies the programming system—this may be Python, Sage, GAP, Mathematica;
3. asks ChatGPT for short, readable code that carries out the core computation.

Once I have a draft, I run it. I correct boundary cases. I adjust the design. The first version is rarely perfect, but it gives me a workable framework.

7.1.2 Choosing Data Structures

ChatGPT is also helpful when I must choose between data representations. For example:

Example

Prompt

Suggest two ways to encode a finite lattice in Python. One option is diagram adjacency lists. Another option is join and meet tables. Discuss the trade-offs for computing joins, meets, and congruences.

Such discussions often surface various issues. These include memory, speed, and ease of checking identities. I might otherwise overlook them.

7.2 Debugging and Refactoring Code

Once I have code, I refine it. This involves correcting mistakes. It also involves reorganizing longer routines into smaller ones. It also involves improving the readability of the entire script.

7.2.1 Explaining Error Messages

Error messages can be obscure, especially in systems with strict type or syntax rules. I often paste a short fragment and ask:

Example

Prompt

Here is my code and the error message it produces. Explain the problem in plain language and suggest a minimal fix. Do not rewrite the whole program.

Even when the diagnosis is not exact, it is usually close enough to lead me to the true issue.

7.2.2 Refactoring for Clarity

Experimental code grows quickly. It accumulates temporary print statements. It accumulates duplicated conditions. It accumulates long functions that do too much. Before sharing the code with coauthors or students, I ask ChatGPT to help me refactor:

Example

Prompt

Rewrite this function to make it clearer and easier to maintain. Break long routines into smaller ones. Improve variable names. Add docstrings. Do not change the external behaviour.

I still test the result, but the improvement in clarity is often substantial.

7.3 Structuring Computational Experiments

Experiments are most useful when they are structured and reproducible: I must know what was computed, under which conditions, and how the results were summarized.

Once again, this follows the standard workflow introduced in the Introduction, with code generation replacing textual drafts.

7.3.1 Planning the Experiment

Before I write code, I often ask ChatGPT to help me plan:

Example

Prompt

Help me design an experiment to test a conjecture about finite modular lattices. What parameters should I vary, what should I record, and how should I organize the output?

Typical suggestions include:

- enumerating structures by size or invariant,
- recording counts of properties and exceptions,
- isolating minimal counterexamples,
- preparing simple plots or LaTeX tables.

Thinking through these issues in advance helps avoid costly reruns.

7.3.2 Turning Output into Readable Tables

After running the code, I often have plain text or log files. Converting this into a clean table is tedious. My workflow is:

1. run the experiment and collect output such as
   `n=6, total=..., modular=..., distributive=...,`
2. paste a representative portion into ChatGPT,
3. ask for a LaTeX table using `\begin{tabular}`.

I verify the numbers, but the formatting work is greatly reduced.

7.4 Union-Indecomposable Sets

In this section, I illustrate how I use ChatGPT to design, implement, and interpret a small computational experiment. The mathematical topic is simple but typical of my work. I focus on sets that cannot be written as nontrivial unions of smaller sets in a finite family.

Recall that if $\mathcal{F}$ is a finite family of subsets of a set X and $A \in \mathcal{F}$, then A is called *union-indecomposable* if

$$A = B \cup C \quad \text{implies that} \quad A = B \text{ or } A = C$$

for all $B, C \in \mathcal{F}$.

7.4.1 The Motivating Question

Example

Prompt

In simple families of finite set systems, such as the family of all subsets of $\{1, \ldots, n\}$, how can we detect union-indecomposable elements automatically, and what patterns do we see in the results?

The goal is not to prove a theorem, but to show how ChatGPT can help set up and explain a small computational exploration.

7.4.2 ChatGPT's Code Candidate

I begin by giving ChatGPT a precise mathematical description, in words.

Example

Prompt

Consider the family of all subsets of $\{1, \ldots, n\}$. The union of two subsets is their set-theoretic union. Write Python code that represents this family and a function `is_union_indecomposable(F, A)` that tests whether a given subset A is union-indecomposable.

(continued)

Candidate

ChatGPT then returns a draft of the code. I do *not* trust it blindly, but I find it useful as a first pass. A typical skeleton looks like this:

Python skeleton (from ChatGPT).

```
from itertools import combinations

def all_subsets(n):
    """Return the list of~all subsets
    of~{1,...,n} as~frozensets."""
    base = list(range(1, n+1))
    F = []
    for~k in~range(n+1):
        for~subset in~combinations(base, k):
            F.append(frozenset(subset))
    return F

def is_union_indecomposable(F, A):
    """Test whether A cannot be~written
    as~a~union of~two smaller subsets."""
    if len(A) == 0:
        return False
    for~B in~F:
        for~C in~F:
            if B.issubset(A) and C.issubset(A):
                if B | C == A and B != A
                and C != A:
                    return False
    return True
```

There are many possible variants; this one is simple enough for exposition. Notice the important steps:

- The elements of the family are represented as `frozensets`.
- The basic operation is set-theoretic union.
- Only subsets of A are considered in possible decompositions.
- The definition of union-indecomposability is translated literally.

7.4.3 Verification and Experiment

Verification
My job, at this stage, is to check that the translation is faithful. For example, I verify by hand that the test `len(A) == 0` correctly excludes the empty set, which is trivially decomposable. I also verify that the condition `B != A` and `C != A` rules out trivial unions.

Once I am satisfied with the basic functions, I ask ChatGPT to extend the code. I want a loop over all subsets. I want it to collect the union-indecomposable ones. I want it to print a summary for several small values of n:

```
def union_indecomposables(F):
    return [A for~A in~F if
    is_union_indecomposable(F, A)]

for n in~range(1, 6):
    F = all_subsets(n)
    U = union_indecomposables(F)
    print(f"n = {n}")
    print(f"  size of~family: {len(F)}")
    print(f"  number of~union-indecomposables:
    {len(U)}")
```

On my own machine, running this code produces output of the form

```
n = 1
  size of~family: 2
  number of~union-indecomposables: 1
n = 2
  size of~family: 4
  number of~union-indecomposables: 2
n = 3
  size of~family: 8
  number of~union-indecomposables: 3
n = 4
  size of~family: 16
  number of~union-indecomposables: 4
n = 5
  size of~family: 32
  number of~union-indecomposables: 5
```

It is often more readable to present this as a small table. ChatGPT can also generate the LaTeX code for such a table:

n	size of the family	number of union-indecomposables
1	2	1
2	4	2
3	8	3
4	16	4
5	32	5

In this very simple family, the pattern is obvious: among all subsets of $\{1, \ldots, n\}$, the union-indecomposable ones are exactly the singleton sets.

7.4.4 Lesson

From a mathematical point of view, nothing deep has happened. We already knew that a finite set can be written as a nontrivial union if and only if it has more than one element.

The point of the example is different:

- We began with a clear mathematical definition (union-indecomposable subsets).
- With the help of ChatGPT, we translated this definition into executable code.
- We ran the code on a small but nontrivial family of examples.
- We summarized the results in a table and interpreted the pattern.

In more complicated situations—for instance, families of sets with additional constraints or operations—the same workflow helps explore patterns, generate conjectures, and locate counterexamples. The computer does not prove theorems, but it helps organize evidence.

Example

Worked example (computation)

ChatGPT translated a set-theoretic definition into executable Python code. It produced a loop to collect statistics for small n and suggested a table presentation. The *mathematical content*—which sets are union–indecomposable and why—remains under human control, but the mechanical work of coding and formatting is accelerated.

7.4.5 What ChatGPT Contributed

In this worked example, ChatGPT assisted in three ways:

1. *Code drafting.* It produced an initial version of the functions implementing the set family and the definition.
2. *Formatting and presentation.* It helped turn raw output into a readable table and supplied LaTeX code.
3. *Explanation.* It suggested language for describing the experiment and its outcome.

In each of these roles, ChatGPT is an *assistant*, not a source of truth. I still need to:

- check that the code matches the mathematical definition,
- run the code myself and verify the output,
- interpret the results and decide what they mean.

This small example illustrates the general theme of this chapter: ChatGPT can help design and explain computational experiments, but the responsibility for correctness, interpretation, and mathematical insight remains with the human mathematician.

7.5 Numerical and Symbolic Computation

Not all experiments are combinatorial. Some are numerical. Others are symbolic.

7.5.1 Designing Numerical Experiments

Numerical experiments can mislead if I choose poor ranges or if rounding corrupts the data. I use ChatGPT as a sounding board:

Example

Prompt

I want to study the growth of a sequence a_n defined by [formula]. Suggest a numerical experiment that would distinguish between n^2, $n^2 \log n$, and n^3, and comment on numerical stability.

Typical suggestions include log log plots. They include checking multiple ranges. They include fitting simple functions.

7.5.2 Guessing Formulas from Data

When I compute the first several terms of a sequence, I often want to guess a closed form. My prompt is usually:

Example

Prompt

> Here are the first ten values of a sequence. Suggest standard combinatorial sequences that match or resemble it. Indicate where to verify the match (e.g., OEIS).

I then check the suggested matches rigorously.

7.6 Limitations and Good Practices

7.6.1 ChatGPT Does Not Run My Code

ChatGPT cannot execute code or observe my environment. After receiving a suggested function, I must copy it into my system. I must run it on test cases. I must verify performance and correctness.

► **Warning** Code suggested by ChatGPT is always a *candidate implementation*. Until I run it, test it, and compare it with the intended mathematical definition, it should not be treated as reliable or publication-ready.

7.6.2 Reproducibility

Computational work is useful only if it is reproducible. I am careful to document:

- code and data versions;
- random seeds or sampling procedures;
- instructions needed to rerun the experiment.

ChatGPT can help me draft these instructions, but the responsibility is mine.

7.7 Prompt Patterns

Here are prompt templates I use frequently in computational work.

7.7.1 Code Design

Example

Prompt

I want to run the following experiment: [description]. Write short, clear code in [Python/Sage/GAP] to carry out the core computation, with enough comments for me to modify it.

7.7.2 Debugging

Examplc

Prompt

Here is a code fragment and its error message. Explain the problem and suggest a minimal fix, without rewriting the whole program.

7.7.3 Experiment Planning

Example
Prompt

Help me plan a computational experiment to test this conjecture. Suggest what to vary. Suggest what to record. Suggest how to summarize the results.

7.7.4 From Output to Exposition

Example

Prompt

Here is my experimental output. Summarize the observed pattern. Suggest a conjecture that would explain it.

7.7.5 Math-Code Bridge

Example

Prompt

Translate this mathematical definition into a function in [system]. Describe corner cases explicitly. Write a docstring explaining the underlying definition.

In this chapter I have focused on computation and experimentation as tools for mathematical exploration. ChatGPT can help design small computational experiments, write and debug code, and summarize the results of systematic searches. Used carefully, these experiments provide evidence, reveal patterns, and uncover counterexamples, but they never replace proof or mathematical judgment.

The next chapter turns from computation to visualization. Diagrams play a central role in many areas of mathematics, from lattice theory and category theory to graphs and probability. Chapter 8 examines how ChatGPT can assist in producing and refining diagrams, and how to recognize when a diagram clarifies structure—and when it misleads.

7.8 Exercises

Exercise 7.1 Choose a small computational task from your own work (for example, enumerating substructures or testing an algebraic law). Ask ChatGPT to outline an algorithm before writing any code. Check whether the algorithm faithfully reflects the mathematics.

Exercise 7.2 Ask ChatGPT to write code for a simple mathematical experiment discussed in this chapter. Run the code yourself and verify that the output matches the expected mathematical results.

Exercise 7.3 Give ChatGPT a piece of code that produces incorrect output. Ask it to diagnose the error. Determine whether the diagnosis addresses the mathematical logic, the implementation, or both.

Exercise 7.4 Ask ChatGPT to refactor a short program so that the mathematical intent is clearer. Compare the original and revised versions and decide which one better reflects the underlying mathematics.

Exercise 7.5 Ask ChatGPT to modify a computational experiment so that it records intermediate results rather than only final outputs. Analyze whether this makes verification easier.

Exercise 7.6 Ask ChatGPT to interpret the results of a small computational experiment. Check whether the interpretation goes beyond what the data justify.

Exercise 7.7 Ask ChatGPT to suggest additional experiments that would test the robustness of a conjecture suggested by computation. Decide which suggestions are mathematically meaningful.

Exercise 7.8 Repeat one computational experiment from this chapter using a different coding approach or language. Compare the results and identify any discrepancies.

Exercise 7.9 Ask ChatGPT to explain the limitations of computational experiments for a problem discussed in this chapter. Compare its explanation with your own assessment.

Exercise 7.10 Design a prompt that encourages ChatGPT to draw conclusions from computational data that are not logically warranted. Identify where mathematical judgment must intervene.

7.9 Transition

This chapter has focused on computation and experimentation, where mathematical ideas are explored through code and data.

In the next chapter, I turn to visualization. Diagrams provide another way to understand structure, complementing both symbolic reasoning and computation. I show how ChatGPT can assist in constructing and refining diagrams, while the mathematician ensures that they accurately represent the underlying ideas.

Diagrams Beyond Lattices

8

Mathematical diagrams reach far beyond lattice diagrams. We use commutative diagrams in algebra. We use graphs in combinatorics. We use trees in logic and computer science. We use geometric figures in Euclidean geometry. We use transition diagrams in probability. In this chapter, I illustrate how ChatGPT assists with constructing this broader range of diagrams. I always follow the workflow established earlier. *A prompt produces a candidate diagram. The human mathematician then verifies and refines it.*

8.1 Why Diagrams Matter

A well-designed diagram can convey structure in a way that text alone cannot. It shows the architecture of a proof. It shows the symmetry of a construction. It shows the hierarchy in a classification.

Example

Prompt → Candidate → Verification

- **Prompt** Describe clearly what the diagram should contain: nodes, edges, labels, placement, style.
- **Candidate** ChatGPT produces TikZ code matching the request.
- **Verification** I refine spacing. I adjust geometry. I check that the figure communicates the intended mathematics.

The more precise the description, the closer the candidate diagram is to what I need.

G. Grätzer, *ChatGPT for Math*, https://doi.org/10.1007/978-3-032-27286-7_8

8.2 Commutative Diagrams

The `tikz-cd` package provides a compact syntax for commutative diagrams, and ChatGPT handles it consistently well.

Example
A basic commuting square

Prompt

Draw a commutative square with maps f, g, h, and k.

Candidate:

```
\begin{tikzcd}
A \arrow[r,"f"] \arrow[d,"g"'] &
B \arrow[d,"h"] \\
C \arrow[r,"k"'] &
D
\end{tikzcd}
```

This is the diagram:

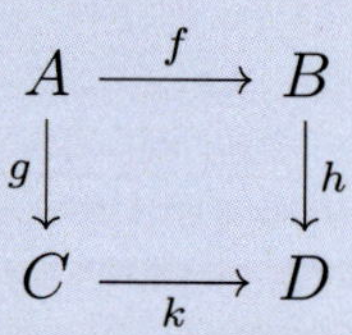

Verification Spacing is adequate. Labels are legible. No revision needed.

I can request diagonal morphisms. I can request dashed arrows. I can request adjusted spacing. ChatGPT handles these reliably.

8.3 Graphs and Digraphs

TikZ handles graphs beautifully, most of the time.

Example

The Petersen graph

(continued)

Prompt

Draw the Petersen graph.

```
\begin{center}
\begin{tikzpicture}[scale=2,
  every node/.style={circle, draw,
  inner sep=1.2pt},
  thick
]
% --- outer 5-cycle (pentagon) ---
\foreach \i/\ang in {0/90,1/18,2/306,3/234,4/162}{
  \node (v\i) at (\ang:1.2) {};
}
% --- inner 5-cycle drawn as~a~star ---
\foreach \i/\ang in {0/90,1/18,2/306,3/234,4/162}{
  \node (u\i) at (\ang:0.6) {};
}
% outer cycle edges
\draw (v0)--(v1)--(v2)--(v3)--(v4)--cycle;
% spokes
\foreach \i in {0,1,2,3,4}{
  \draw (v\i)--(u\i);
}
% inner star edges (u0-u2-u4-u1-u3-u0)
\draw (u0)--(u2)--(u4)--(u1)--(u3)--cycle;
\end{tikzpicture}
\end{center}
```

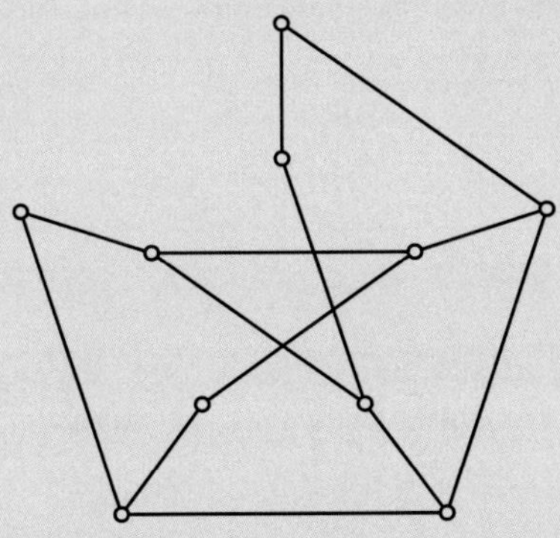

Verification There should be 15 edges and each vertex should have degree 3.

Ouch. Tell ChatGPT that many edges are missing.

8.4 Trees and Hierarchical Structures

Trees appear in many contexts. ChatGPT is particularly strong at producing clean hierarchical layouts.

Example

A rooted tree

Prompt

Draw a rooted tree with one root, three children, and two leaves under the middle child.

Candidate

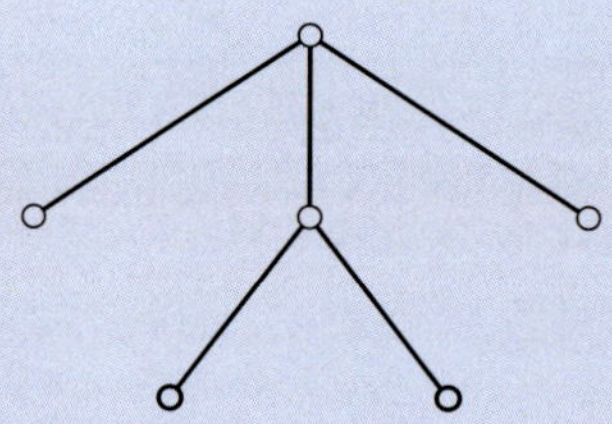

ChatGPT returns a valid layered tree with nodes on two horizontal levels and arrows pointing downward.

Verification

The structure is correct. Size and spacing of the leaves may need small adjustments for publication.

8.5 Geometric Figures

TikZ is excellent for precise geometric illustrations. It can produce candidate code for triangles, for circles, for angle markings, for coordinate grids, and for shaded regions.

Example
Triangle with labeled vertices

Prompt

Draw triangle ABC and label its vertices, circles red, edges blue. TikZ.

Candidate

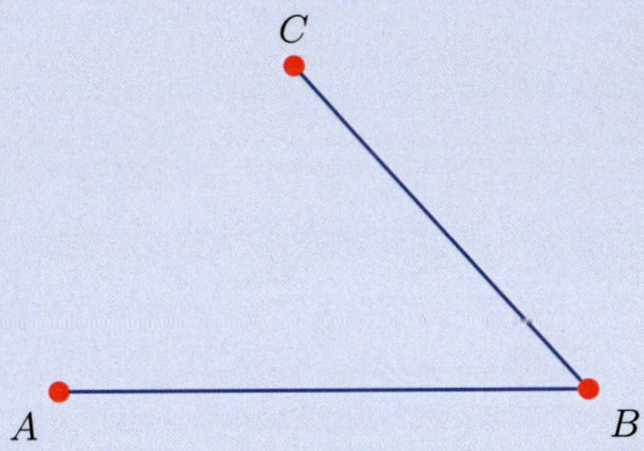

Verification

An edge is missing.

Tell ChatGPT that an edge is missing.

► **Warning**
Beware of misleading geometry. Even if a candidate diagram is mathematically correct, its *visual* presentation may mislead. Angles may appear equal when they should not. Circles may look distorted. Parallel lines may converge. Always check that the geometry communicates the intended structure.

8.6 Probability and Markov chain diagrams

Diagrams with states and weighted transitions are routine in probability theory. ChatGPT can generate clean candidates with labeled arrows and loops.

Example
A three-state Markov chain

Prompt

Draw a Markov chain with states A, B, C in a row and labeled transitions $A \to B$, $B \to C$, and $C \to A$.

Candidate

Circular nodes in a line with directed arrows and labels.

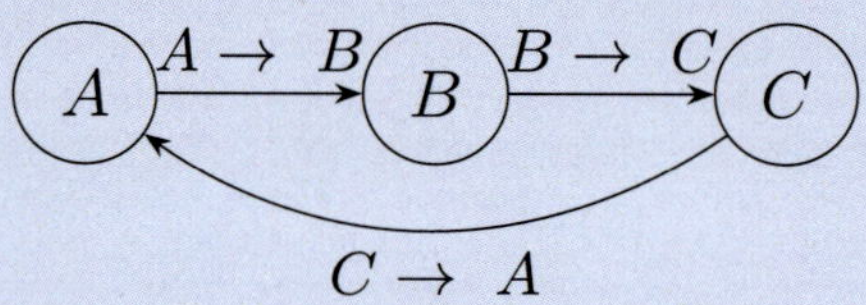

Verification

The geometry and labels are correct. I typically adjust spacing so arrow labels do not touch the edges.

Larger chains or branching structures require more detailed refinement. This includes loop curvature, consistent arrow angles, evenly spaced states.

8.7 Category-Theoretic Diagrams

Category theory relies heavily on diagrammatic reasoning. The `tikz-cd` syntax aligns perfectly with ChatGPT's strengths.

Example
Pullback square with universal property

Prompt

Draw the pullback of $f\colon X \to Z$ and $g\colon Y \to Z$ and include the mediating morphism $u\colon W \to P$.

(continued)

Candidate

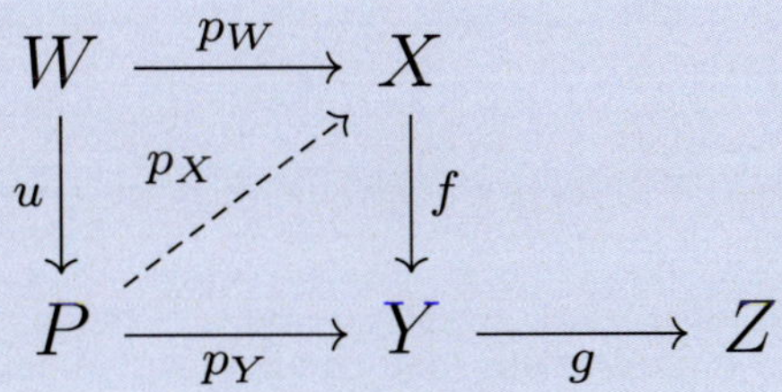

ChatGPT produces the correct commuting diagram with π_1, π_2, and a dashed arrow from P to W.

Verification

All arrows have correct directions and labels. For clarity, I increase row and column spacing before publication.

The correctness of categorical diagrams is a matter of syntax rather than geometry, making ChatGPT particularly effective in this area.

8.8 The Human Touch

Even the best candidate diagrams need refinement. Arrows may meet at shallow angles. Labels may be crowded. Spacing may be suboptimal.

► **Warning**

ChatGPT accelerates drafting, but the mathematician must verify:

- the diagram truly represents the intended mathematics,
- the geometry does not mislead,
- spacing, angles, and labels meet publication standards.

A candidate is only a starting point.

Example
Human refinement improves clarity ChatGPT often produces a commuting square with compact spacing. Increasing row and column separation by hand dramatically improves readability while preserving structure.

The iterative workflow—*prompt* → *candidate* → *refinement*—is the core methodology throughout this chapter as well as the previous ones.

8.9 Exercises

Exercise 8.1 Ask ChatGPT to solve a nontrivial theorem–proof problem from your own area of expertise.

1. Identify at least two points where mathematical judgment, rather than symbol manipulation, was required.
2. Explain why these points cannot be reliably delegated to a language model.

Exercise 8.2 Prompt ChatGPT to give a confident proof of a true but subtle theorem.

1. Verify the proof line by line.
2. Locate any unjustified inference, hidden assumption, or incorrect claim.
3. Rewrite the proof, marking explicitly which steps required human verification.

Exercise 8.3 Provide ChatGPT with a flawed argument and ask it to "fix" the proof.

1. Compare the repaired version with a correct proof from the literature.
2. Identify whether the repair was genuinely mathematical or merely rhetorical.

Exercise 8.4 Design a prompt that subtly encourages ChatGPT to guess (for example, by asking for an obscure theorem or attribution).

1. Record the model's response.
2. Classify any hallucination you observe: invented fact, invented citation, or invented reasoning.
3. Explain how the standard workflow would detect this failure.

Exercise 8.5 Ask ChatGPT to generalize a correct result beyond its natural scope.

1. Identify where the generalization fails.
2. Explain why plausibility is not evidence.

Exercise 8.6 Have ChatGPT generate a diagram meant to support a mathematical claim.

1. Determine whether the diagram encodes assumptions not stated in the text.
2. Decide whether the diagram clarifies or obscures the argument.

Exercise 8.7 Imagine you are reviewing a paper whose proofs were heavily assisted by ChatGPT.

1. What verification steps would you require?
2. What disclosure, if any, would you expect from the author?

Exercise 8.8 List three mathematical tasks from your own work where using ChatGPT would introduce unacceptable risk. For each, explain why the risk cannot be mitigated by prompting or verification.

Exercise 8.9 Modify the standard workflow from the Introduction to emphasize limits rather than power.

1. Which steps become more prominent?
2. Which steps should be slowed down or repeated?

Exercise 8.10 Write a short policy for your own mathematical work answering:

> When do I use ChatGPT, when do I avoid it, and what responsibilities do I retain regardless?

8.10 Transition

Diagrams help us externalize structure and communicate mathematical ideas visually. ChatGPT accelerates their production, but accuracy and interpretation remain human responsibilities.

In the next chapter, I turn from visualization to teaching. I examine how ChatGPT can assist in preparing explanations, exercises, and instructional materials, while the instructor retains control over mathematical content and pedagogical goals.

Teaching with ChatGPT

9

I teach as much as I do research. I prepare lectures, homework, quizzes, exams, examples, counterexamples. I prepare explanations at many different levels. In this chapter, I illustrate how ChatGPT assists with teaching. I follow the same workflow used throughout this book. *A prompt produces a candidate teaching artifact. The instructor then verifies and refines it.*

9.1 ChatGPT and the Teaching Workflow

Teaching is a steady stream of small, concrete tasks. ChatGPT is well suited to help with these tasks because it can produce clear, mathematically literate drafts quickly.

Example

Prompt $\rightarrow$ Candidate $\rightarrow$ **Verification**

- **Prompt**

  State the topic. State the level. State the constraints. These include prerequisites, notation, time limits, and the intended learning goal.

- **Candidate**

  ChatGPT produces a draft. This may be an explanation, a problem set, a quiz, or a worked example.

- **Verification**

  I check correctness, adjust difficulty and tone. I align the result with my course and my standards.

G. Grätzer, *ChatGPT for Math*, https://doi.org/10.1007/978-3-032-27286-7_9

The key point is methodological: ChatGPT accelerates drafting, but it does not replace judgment.

9.2 Multiple Levels of Explanation

One of the most useful ways I work with ChatGPT is by asking for the same mathematical idea to be explained at different levels. This is especially effective when a single diagram supports several explanations, each aimed at a different audience.

Example

A directed cycle, three explanations

Prompt
Explain the same diagram at three levels: informal, standard, and domain-specific.

Example

Diagram

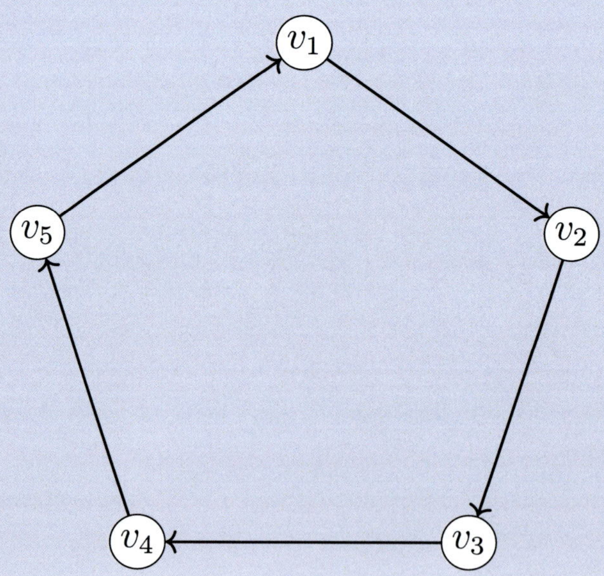

(continued)

Candidate

Level 1 informal explanation.
This picture shows five points arranged in a circle. Each arrow points to the next point. If you keep following the arrows, you eventually come back to where you started.

Level 2 standard mathematical explanation.
The diagram represents a directed cycle of length 5. It consists of five vertices $v_1, \ldots, v_5$ and directed edges $v_i \rightarrow v_{i+1}$ (with indices taken modulo 5). Such cycles appear naturally in graph theory. They appear in dynamical systems. They appear in the study of finite automata.

Level 3 domain-specific explanation.
Viewed as a Markov chain with deterministic transitions, this diagram represents a system whose state advances cyclically through five positions. The transition matrix has exactly one nonzero entry in each row, and the chain is periodic with period 5.

Verification
The mathematics is correct. What I change is pedagogical. I adjust terminology, pacing, and emphasis. The diagram itself does not change. The language does.

This ability to generate parallel explanations is a genuine advantage in teaching and writing, but I always check accuracy and suitability for the intended audience.

9.3 Homework Sets

One of the most immediate and useful applications of ChatGPT is the generation of homework problems. Given a topic and a desired level of difficulty, it can produce varied drafts suitable for practice, discussion sections, or informal assessment. The instructor's job is to curate.

Example

From topic to a usable homework set

Prompt

> Generate five homework problems on equivalence relations for an introductory course, ranging from routine verification to short proofs.

(continued)

Candidate

A typical mix includes:

- verifying that a given relation is an equivalence relation,
- finding equivalence classes on a finite set,
- proving that equivalence classes form a partition,
- constructing an equivalence relation from a given partition,
- identifying common mistakes in proposed relations.

Verification

I tighten statements, check edge cases, and adjust difficulty. If solutions are requested, I verify them independently.

Example

Building a long-term problem bank

Prompt

> Generate twenty problems on homomorphisms of ordered sets at three levels of difficulty and label them accordingly.

Candidate

A labeled bank (routine, intermediate, challenge) that can be stored, edited, and reused across multiple terms.

Verification

I remove duplicates. I fix ambiguous wording. I rewrite the few problems whose "difficulty label" does not match my course.

► **Warning**

Drafts, not finished products

Problems generated by ChatGPT are raw material. Statements may need tightening. Corner cases may be overlooked. Solutions must be checked carefully. Used as a drafting assistant rather than an authority, ChatGPT integrates smoothly into an instructor's workflow.

9.4 Quizzes and Exams

ChatGPT can assist with quizzes and examinations in a direct way. It can produce draft questions. It can produce variants. It can produce structured sets of problems from carefully written prompts. Its strength is speed and breadth. Its limitation is its lack of judgment. For high-stakes assessment, verification matters even more.

Example

A ten-minute quiz draft

Prompt

Create a ten-minute quiz on isotone maps of ordered sets.

Candidate

A typical draft includes:

1. State the definition of an isotone map between ordered sets.
2. Let $P = \{a, b, c\}$ be an ordered set with $a < b$ and $a < c$, and let $Q = \{0, 1\}$ with $0 < 1$. Define $f : P \to Q$ by $f(a) = 0$, $f(b) = 1$, $f(c) = 1$. Is f isotone?
3. Explain briefly why not every map between ordered sets is isotone.

Verification

I check that the ordered-set relations support the intended reasoning and that the quiz fits the time limit.

Example

Parallel versions with the same structure

Prompt

Produce three parallel exam questions on isotone maps of ordered sets, each with the same difficulty but different underlying ordered sets.

Candidate Three variants with the same structure but different labels and data.

(continued)

Verification I ensure the variants are genuinely comparable in difficulty and that no variant accidentally becomes easier. I also ensure no variant accidentally becomes harder. This can happen because of hidden structure.

> **Warning**
> **High-stakes settings**
> Even when questions look correct, subtle issues may remain. These include overlooked order relations. These include unintended counterexamples. These include mismatched difficulty. All quiz and exam questions produced by ChatGPT must be checked carefully.

9.5 Worked Examples

Examples are central to learning mathematics. ChatGPT can generate worked examples in many areas. These include linear algebra, abstract algebra, analysis, and combinatorics. They also include advanced topics such as universal algebra or lattice theory. The value is not that it proves the theorem for you. The value is that it produces a coherent first draft of an explanation.

Example

A worked example of a conditional theorem

Prompt

> Give a worked example illustrating a conditional theorem from real analysis. Show each step, and make clear exactly where the hypothesis is used.

Candidate

A typical candidate uses the Mean Value Theorem. It chooses a concrete function and a concrete interval, verifies the hypotheses, and then carries out the conclusion step by step.

For example, take $f(x) = x^2$ on the interval $[1, 3]$. We verify the hypotheses:

- f is continuous on $[1, 3]$.
- f is differentiable on $(1, 3)$.

(continued)

Therefore the Mean Value Theorem applies, so there exists $c \in (1, 3)$ such that

$$f'(c) = \frac{f(3) - f(1)}{3 - 1}.$$

Compute each term explicitly:

$$f(3) = 9, \qquad f(1) = 1, \qquad \frac{f(3) - f(1)}{3 - 1} = \frac{9 - 1}{2} = 4.$$

Also

$$f'(x) = 2x,$$

so the conclusion becomes

$$2c = 4,$$

hence

$$c = 2.$$

Finally, check that the computed value satisfies the required condition:

$$2 \in (1, 3).$$

Verification

I check that the hypotheses are stated and verified (continuity on $[1, 3]$ and differentiability on $(1, 3)$). I check that the average slope is computed correctly. I check that the derivative is computed correctly. I check that solving $2c = 4$ is done transparently and that the final value indeed lies in the open interval $(1, 3)$.

I sometimes keep the same mathematics but rewrite the presentation. I write one version with every algebraic step explicit. I write a shorter version for students who already have fluency.

9.6 Visual Explanations

Diagrams often clarify a concept faster than paragraphs of text. ChatGPT can produce TikZ code that turns an informal sketch into a clean figure ready for lecture notes, slides, or a book.

Example

Three pictures of a function

Prompt

Draw three small diagrams illustrating injective, surjective, and bijective maps.

Candidate

A typical response is a triple of simple arrow diagrams.

Verification

I adjust spacing and labels and I ensure the picture matches the definition I am using in class (especially when discussing codomain versus image).

Example

Small ordered sets for classroom discussion

Prompt

Draw five small ordered sets illustrating different order properties.

Candidate

A multi-panel figure (chain, antichain, diamond, and other small examples).

Example

Verification

I verify the intended comparabilities, then refine the geometry and styling so the diagrams match the rest of my notes.

The main advantage is not that ChatGPT draws diagrams that I could not draw myself. The main advantage is that it produces a usable first draft of TikZ code without starting from a blank page.

9.7 Lectures and Students

A short prompt can generate a lecture outline that is immediately useful.

Example
From prompt to lecture outline

Prompt

Prepare a 50-minute lecture on the Fundamental Theorem of Finite Abelian Groups. Include: (1) motivation, (2) formal statement, (3) examples, (4) two proofs (sketches), and (5) common student misconceptions.

Candidate

A structured outline with headings, timing suggestions, and standard examples.

Verification

I adjust the pacing to match my class. I replace generic examples with the ones I prefer. I rewrite the proof sketches in my own notation.

I can also paste my own draft notes and ask for improvements. These include clearer transitions, better notation choices, and missing motivation. I can also request a more student-friendly ordering of topics.

9.7.1 Checking Student Work

One of the most natural uses of ChatGPT in teaching is to help check student work. A student submits a solution. I paste it into ChatGPT with a carefully written prompt. I receive a candidate evaluation. This is a diagnosis of errors and a short piece of feedback. This is useful but it must be verified since an incorrect diagnosis can be worse than none.

Example

Induction proof: diagnose, do not rewrite

Prompt

You are a teaching assistant. I will give you an exercise and a student solution. (1) Decide whether it is correct. (2) If not, locate the error precisely and explain it simply. (3) Give constructive feedback in 2–3 sentences. Do not provide a model solution unless asked.

Exercise

Prove by induction that for every integer $n \geq 1$,

$$1 + 2 + \cdots + n = \frac{n(n+1)}{2}.$$

Candidate

ChatGPT typically identifies that the structure is right but that the algebra in the induction step is incorrect. The $(n + 1)$ terms cannot be "cancelled". They must be combined.

Verification

I check that the diagnosis is correct. I edit the feedback so it matches my tone and the expectations of the course.

In large classes, this can function as a first filter. It sorts solutions into "essentially correct". It sorts solutions into "minor algebraic error". It sorts solutions into "serious gap" before I read them myself. As always, the final judgment is mine.

9.8 Exercises

Exercise 9.1 Ask ChatGPT to describe how it expects its mathematical capabilities to improve over the next five years.

1. Identify which claims are plausible extrapolations and which are speculative.
2. Explain why technological optimism is not a substitute for verification.

Exercise 9.2 Select a mathematical task that ChatGPT currently performs poorly.

1. Analyze whether this limitation is conceptual, architectural, or data-driven.
2. Decide whether the limitation is likely to disappear or to persist.

Exercise 9.3 Ask ChatGPT to redesign the standard workflow assuming a future model with significantly stronger reasoning abilities.

1. Identify which safeguards remain essential.
2. Identify which safeguards, if any, might reasonably be relaxed.

Exercise 9.4 Consider a classroom or research group that routinely uses ChatGPT.

1. Propose rules or norms that would preserve mathematical integrity.
2. Identify failure modes that institutional rules cannot prevent.

Exercise 9.5 Ask ChatGPT to propose a fully automated system for producing mathematical papers.

1. Analyze where responsibility becomes unclear or diffused.
2. Explain why accountability cannot be automated.

Exercise 9.6 Identify a historical episode in mathematics where a powerful new tool changed practice (for example, symbolic algebra systems or computer-assisted proofs).

1. Compare that transition with the introduction of ChatGPT.
2. Identify similarities and crucial differences.

Exercise 9.7 Ask ChatGPT to evaluate its own reliability as a long-term mathematical assistant.

1. Identify where the self-assessment is informative.
2. Identify where it becomes circular or vacuous.

Exercise 9.8 Imagine a future reader encountering a theorem whose proof was produced with substantial assistance from ChatGPT.

1. What information would that reader need to trust the result?
2. What information cannot be conveyed after the fact?

Exercise 9.9 Identify one way in which reliance on ChatGPT could subtly change how mathematics is written, taught, or valued.

1. Decide whether this change would be beneficial or harmful.
2. Justify your answer.

Exercise 9.10 Write a brief statement describing how you intend to use tools like ChatGPT five years from now. Address not only what you hope such tools will do, but what you insist on continuing to do yourself.

9.9 Transition

Teaching highlights how ChatGPT can support explanation, assessment, and feedback, while leaving judgment and responsibility with the instructor.

In the next chapter, I step back and make this interaction explicit. I present a collection of prompt patterns that capture the workflows developed so far, providing a practical toolkit for using ChatGPT effectively in mathematical work.

10 Prompt Patterns for Math

In this chapter, I collect a number of reusable “prompt patterns” for working with ChatGPT. Each pattern consists of a short description of its use. Each pattern includes a template prompt that I (or a student) can adapt. Each pattern includes a brief example from math.

My goal is to make it easier for me to start an interaction in a controlled way, rather than improvising every prompt from scratch.

10.1 Explain, then Test Me

When I learn a new concept or revise an old one, it is helpful to first see an explanation. Then I answer questions about it.

Example

Template prompt

> Explain the concept of X at the level of a strong math undergraduate. Then ask me three short questions to check my understanding. After each answer, tell me whether it is correct. If it is not correct, give a model answer. Do not move on to the next question until I have answered the current one.

G. Grätzer, *ChatGPT for Math*, https://doi.org/10.1007/978-3-032-27286-7_10

Specifically,

Example

Prompt
Explain the Mean Value Theorem at the level of a strong math undergraduate. Then ask me three short questions to check my understanding.

Worked example

Question 1. State the Mean Value Theorem.

Answer

Let $f : [a, b] \to \mathbb{R}$ be a function that is continuous on $[a, b]$ and differentiable on (a, b). Then there exists a point $c \in (a, b)$ such that

$$f'(c) = \frac{f(b) - f(a)}{b - a}.$$

Question 2. Give an example where the Mean Value Theorem applies.

Answer

For $f(x) = x^2$ on the interval $[1, 3]$, the function is continuous on $[1, 3]$ and differentiable on $(1, 3)$. Therefore the Mean Value Theorem applies, and there exists $c \in (1, 3)$ such that

$$f'(c) = \frac{9 - 1}{2} = 4.$$

Since $f'(x) = 2x$, this gives $c = 2$.

Question 3. Give an example of a function and interval for which the Mean Value Theorem does *not* apply.

Answer. The function $f(x) = |x|$ on the interval $[-1, 1]$ does not satisfy the hypotheses of the Mean Value Theorem, because it is not differentiable at $x = 0$.

Verification
This pattern encourages active engagement. Instead of passively reading a theorem, the student must recall its statement. The student must recognize when the hypotheses are satisfied. The student must also identify when and why the theorem fails. Immediate feedback reinforces correct understanding.

10.2 Multi-level Explanations

As a mathematician, I often need explanations at different levels.

Example

Template prompt

> Explain the concept of X at three levels. Write one explanation for a strong math undergraduate. Write one explanation for a first year graduate student. Write one explanation for an expert in another branch of math. Keep each explanation to at most one paragraph.

Specifically,

Example

Prompt

> Explain the concept of a congruence on a lattice at three levels.

Strong math undergraduate

A congruence on a lattice is an equivalence relation that respects the lattice operations.

First year graduate student

A congruence on a lattice L is the kernel of a lattice homomorphism. The set of all congruences of L forms a lattice $\mathrm{Con}\, L$.

Expert in another branch of mathematics

A congruence on a lattice is the appropriate notion of quotient structure.

Verification
The mathematics is the same in the three explanations. The assumed background changes and only the vocabulary alters the emphasis.

Example

Prompt

> Explain the concept of a quotient space in topology at three levels.

Strong math undergraduate

A quotient space is a topological space obtained by identifying certain points of an existing space and giving the result the "largest" topology that makes the identification map continuous.

First year graduate student

Let X be a topological space and let $\sim$ be an equivalence relation on X. The quotient space $X/\sim$ is the set of equivalence classes equipped with the quotient topology: a set $U \subseteq X/\sim$ is open if and only if its preimage under the canonical projection $\pi : X \to X/\sim$ is open in X. This topology is the finest one for which π is continuous.

Expert in another branch of mathematics

A quotient space is the universal way to impose identifications on a space: it is characterized by the property that continuous maps out of the quotient are exactly the continuous maps out of the original space that respect the identifications.

Verification
The underlying mathematics is the same in all three explanations. Only the vocabulary changes. What changes is emphasis and assumed background: from concrete identifications, to formal construction, to universal property.

10.3 Alternative Proof

Given a lemma or theorem with one known proof, I may suspect another exists.

Example

Template prompt

Here is a lemma and a proof.
Give me a different proof that avoids a specified technique. Use an alternative viewpoint. Do not simply paraphrase the original argument.

Specifically,

Example

Prompt

Reprove the representation of a finite ordered set using its indecomposable elements.

Let $(P, \leq)$ be a finite ordered set. Call an element $j \in P$ *indecomposable* if it cannot be written as the maximum of two strictly smaller elements. Let $J(P)$ denote the set of all indecomposable elements of P.

For each $x \in P$, define

$$\varphi(x) = \{\, j \in J(P) \mid j \leq x \,\}.$$

Then $\varphi(x)$ is a down-set of $J(P)$.

The map $\varphi\colon P \to \mathcal{D}(J(P))$, sending each element to the set of indecomposables below it, is order-preserving and injective. Moreover, every down-set of $J(P)$ arises in this way. Thus P is represented faithfully as a family of down-sets of $J(P)$, ordered by inclusion.

Verification
The proposed proof must be checked carefully, but it can reveal connections worth emphasizing.

10.4 Error Doctor

Error messages can be cryptic.

Example

Template prompt

Explain in plain language what the error means. List likely causes. Tell me where to look first.

Example

Worked example

```
! Undefined control sequence.
l.106 \example
```

Verification

The command `\example` is undefined. I inspect the preamble to see whether it was meant to be a macro or an environment.

► Warning

Do not trust diagnoses blindly

Error explanations can be incomplete or wrong. An incorrect diagnosis can waste more time than no diagnosis at all.

10.5 Summarize and Compare

When I read several related papers, I want a quick map of the territory.

Example

Template prompt

Summarize each paper briefly, then compare results and methods.

Example

Worked example

Paper A gives a strong positive result.
Paper B extends the setting.
Paper C provides limiting counterexamples.

Verification
The comparison provides orientation but does not replace reading the papers.

10.6 Structured Rewriting

Often I have a rough paragraph that needs polishing.

Example

Template prompt

Rewrite the paragraph, preserving all mathematics. Remove unnecessary intensifiers.

Example

Worked example

Clearly, it is obvious that the group G is abelian.

Rewritten

The group G is abelian.

Verification
No mathematical content was changed. Only redundant phrasing was removed.

10.7 Exercises

Exercise 10.1 Review a piece of your own mathematical writing composed before you used ChatGPT.

1. Identify which parts could now be assisted by ChatGPT.
2. Identify which parts should remain entirely human-driven.

Exercise 10.2 Ask ChatGPT to summarize the main lessons of this book.

1. Identify which points reflect genuine understanding.
2. Identify which points are generic or superficial.

Exercise 10.3 Revisit the standard workflow introduced earlier in the book.

1. Identify one step you initially underestimated.
2. Explain how your view of that step has changed.

Exercise 10.4 Imagine a junior mathematician relying heavily on ChatGPT.

1. Identify one benefit and one long-term risk.
2. Explain how mentoring could mitigate the risk.

Exercise 10.5 Consider a result whose correctness you personally guarantee.

1. Identify which parts of the work could safely involve ChatGPT.
2. Identify which parts you would never delegate.

Exercise 10.6 Ask ChatGPT to explain why verification is essential in mathematics.

1. Compare its explanation with your own.
2. Identify what the model cannot adequately express.

Exercise 10.7 Identify a mathematical habit that you intend to strengthen as a result of this book. Explain how the use of ChatGPT could either support or undermine that habit.

Exercise 10.8 Write a brief statement answering the following question:

> What responsibilities do I retain even when I use powerful automated tools?

10.8 Transition

The previous chapters have focused on workflows: how ChatGPT can assist with proofs, examples, computation, teaching, and prompting.

We now turn to the infrastructure that supports mathematical work. In the next part, I examine how ChatGPT can assist with bibliographies, including generating, cleaning, and verifying bibliographic data—tasks that are essential in any substantial mathematical project.

Part III

Infrastructure and Writing

Bibliographies

11

Every mathematical paper contains a bibliography or list of references. Many books also contain an index. These tools serve the reader, but preparing them can be tedious. ChatGPT can accelerate both tasks. In this chapter, I describe workflows for generating bibliographic data, for cleaning bibliographic data, and for searching the mathematical literature. I also describe workflows for correcting and completing BibTEX entries.

Although the subject matter is infrastructural, the interactions in this chapter follow the same standard workflow introduced in the Introduction.

The next chapter deals with an index.

11.1 Generating BibTEX Entries

ChatGPT can produce clean BibTEX entries for most mathematical publications. For example:

Example

Prompt

Give me a BibTEX entry for Birkhoff's *Lattice Theory*.

```
@book{Birkhoff1967,
  author    = {Birkhoff, Garrett},
  title     = {Lattice Theory},
  edition   = {Third},
  series    = {Colloquium Publications},
  volume    = {25},
```

(continued)

G. Grätzer, *ChatGPT for Math*, https://doi.org/10.1007/978-3-032-27286-7_11

```
    publisher = {American Mathematical Society},
    address   = {Providence, RI},
    year      = {1967}
}
```

ChatGPT produced

- correct fields (author, title, year, edition),
- a reasonable key,
- consistent formatting.

Here is another approach.

Example

Prompt

Create a BibTEX entry for a specific paper using the DOI. Verify the metadata before adding it to the database.

Input (what I know)

- DOI: `10.1023/A:1005855124769`
- The paper is in *Order* (journal).
- Authors include Grätzer and Wang.

Candidate BibTEX

```
@article{GratzerWang1997LowerBound,
  author  = {Gr{\"a}tzer, George and Wang, Dabin},
  title   = {A lower bound for~congruence representations},
  journal = {Order},
  volume  = {14},
  number  = {1},
  year    = {1997},
  pages   = {67--74},
  doi     = {10.1023/A:1005855124769}
}
```

► **Warning**

Do not trust the candidate entry yet

Even when the DOI is correct, ChatGPT often gets one of these entries wrong. It may get

- the page range wrong,
- the issue number wrong,
- author accents wrong,
- the capitalization wrong,
- the journal's exact subtitle or ISSN fields wrong.

Always verify.

Verification Checklist (What I Verify Before Keeping It)

1. Authors: spelling, order, accents (e.g., `Grätzer`).
2. Title: exact wording and capitalization.
3. Journal metadata: *Order*, volume, issue or number, year.
4. Pagination: first page and last page.
5. DOI: matches the paper (not a nearby erratum or editorial).

Note If you have an MR number, prefer it as the BibTeX key. It makes deduplication easier. It makes later checking easier.

11.2 Cleaning and Deduplicating Bibliographies

Large projects inevitably accumulate messy or duplicated bibliographies. Entries may be copied from different sources. Capitalization may be inconsistent. Fields may be missing. A reliable workflow is:

> *Here are five BibTeX entries. Remove duplicates. Standardize formatting. Complete missing fields. Produce a clean bibliography.*

11.2.1 Worked Example 1: Detecting and Merging Duplicate Entries

Example

Goal

Detect duplicate BibTeX entries that refer to the same paper. Merge them into a single canonical entry.

Example

Prompt

Detect duplicate BibTEX entries:

```
@article{GratzerWang1997,
  author  = {G. Gr\"atzer and D. Wang},
  title   = {A lower bound
  for~congruence representations},
  journal = {Order},
  volume  = {14},
  year    = {1997},
  pages   = {67--74}
}

@article{MR1468957,
  author  = {Gr{\"a}tzer, George and Wang, Dabin},
  title   = {A Lower Bound
  for~Congruence Representations},
  journal = {Order},
  volume  = {14},
  number  = {1},
  year    = {1997},
  pages   = {67--74},
  doi     = {10.1023/A:1005855124769}
}
```

Result

- The entries describe the same article (same authors, title, journal, year, volume, pages).
- Differences are cosmetic (author initials versus full names, title capitalization, missing issue and DOI).

► **Warning**

Common failure mode

ChatGPT may incorrectly merge two distinct papers by the same authors from the same year. Always check the title and pagination.

11.3 Completing Bibliographic Information

Often we remember an author and a year, but not the publication details. In many cases, ChatGPT can reconstruct part of the missing information with surprising accuracy.

Example

Prompt

Complete this entry
J. Smith, "A theorem on ordered sets," 1998.

Result

```
@misc{Smith1998TheoremOrderedSets,
  author = {Smith, J.},
  title  = {A theorem on~ordered sets},
  year   = {1998},
  note   = {Incomplete reference: publication
  venue,
volume/issue, and pages unknown.}
}
```

Even after cleaning and deduplication, bibliographic entries are often *incomplete*. Typical problems include missing page ranges, absent DOIs, incomplete author names, and unspecified publication details (publisher, city, series).

ChatGPT is particularly effective at completing such entries, provided that the interaction is structured and that every added field is independently verified.

Example

Prompt Complete this book entry by adding publisher, place of publication, edition, and year.

```
@book{GratzerLatticeTheory,
  author = {Gr{\"a}tzer, George},
  title  = {General Lattice Theory}
}
```

(continued)

Candidate entry

```
@book{GratzerGLT1978,
  author    = {Gr{\"a}tzer, George},
  title     = {General Lattice Theory},
  publisher = {Birkh{\"a}user},
  address   = {Basel},
  year      = {1978}
}
```

▶ Warning
Edition ambiguity

Many classic books exist in multiple editions. ChatGPT may silently choose the wrong one. Always decide explicitly which edition you intend to cite.

Final Decision (Explicit Edition)

```
@book{GratzerGLT1998,
  author    = {Gr{\"a}tzer, George},
  title     = {General Lattice Theory},
  edition   = {2},
  publisher = {Birkh{\"a}user},
  address   = {Basel},
  year      = {1998}
}
```

A safe principle is to add missing fields conservatively.

Example

Prompt Improve the following entry without guessing. Add only fields that are certain. Leave the rest unchanged.

```
@article{SomeAuthor2020,
  author  = {Someauthor, A.},
  title   = {On lattices},
  journal = {Journal of~Algebra},
  year    = {2020}
}
```

(continued)

Outcome

The entry is left unchanged until authoritative metadata is located.

▶ **Warning**

Best practice

An incomplete but correct entry is better than a complete but wrong one. Never fill fields simply to make an entry "look finished".

11.4 Searching the Mathematical Literature

Searching the mathematical literature is not about asking ChatGPT *what is true*. It is about asking it *where to look*. Used correctly, ChatGPT functions as a fast, knowledgeable research assistant. It helps formulate searches, identify likely venues, and generate candidate references that must then be verified.

ChatGPT is effective at:

- suggesting standard terminology and synonyms,
- identifying likely journals or subject areas,
- recalling well-known theorems and names associated with a topic,
- producing candidate citations to be checked.

It is not reliable as a definitive bibliographic authority.

11.4.1 Worked Example 1: Finding the Right Search Terms

Example
Identify standard terminology and keywords for a vague mathematical idea before searching MathSciNet or zbMATH.

Input (Informal Description)

A result about representing finite lattices by congruence relations.

Example

Prompt (template)

Rewrite the following informal description using standard mathematical terminology. List keywords and phrases suitable for searching MathSciNet or zbMATH.

ChatGPT response

- *congruence lattice*
- *congruence representation*
- *finite lattice*
- *representation theorem*

Outcome

These terms lead quickly to relevant MathSciNet or zbMATH. searches, while the original informal phrasing does not.

11.4.2 Worked Example 2: Locating Likely Venues

Example

Goal

Determine where a paper on ordered sets or lattice theory is most likely to be published.

Example

Prompt

List journals where papers on ordered sets and lattice theory commonly appear.

ChatGPT Response (Typical)

- *Order*

- *Algebra Universalis*
- *Journal of Pure and Applied Algebra*
- *Discrete Mathematics*

► **Warning**
Do not infer existence

The fact that a journal is appropriate does not mean that a specific paper actually appeared there.

11.4.3 Worked example 3: generating candidate references

Example

Goal

Generate candidate references to be checked against authoritative databases.

Example

Prompt

List well-known papers related to congruence lattices of finite lattices. For each item, indicate that the reference must be verified.

ChatGPT Response (Excerpt)

- G. Grätzer, "The Congruences of a Finite Lattice" (verify details).
- R. P. Dilworth, papers on lattice representations (verify exact citations).

► **Warning**
Hallucinated precision

ChatGPT often supplies plausible titles or dates that are slightly wrong. Treat every generated citation as a lead, not as a reference.

11.4.4 Best Practice: The Search Workflow

1. Use ChatGPT to refine terminology and generate leads.
2. Search MathSciNet, zbMATH, publisher sites, or arXiv.
3. Verify every bibliographic field independently.
4. Only then add entries to your BibTEX database.

Used this way, ChatGPT shortens the search phase dramatically while leaving scholarly judgment exactly where it belongs: with the author.

11.5 A Modern BibTEX

BIBLATEX is a modern system for managing bibliographic data in LATEX documents. Its purpose is not merely to format references, but to give the author greater control over how bibliographic information is stored, reused, and presented.

At a high level, BIBLATEX separates three concerns that are often entangled in traditional workflows: the bibliographic database itself, the rules for formatting references, and the way citations are used in the text. This separation makes it easier to adapt a bibliography to different publishers, audiences, or conventions without rewriting the underlying data.

BIBLATEX supports richer bibliographic information than older systems. It can record multiple authors or editors with precise name handling, distinguish clearly between publication roles, and accommodate modern reference types such as online resources, preprints, datasets, and software. It also handles multilingual bibliographies more naturally, allowing titles, notes, and publication details to appear in different languages when needed.

For long documents, BIBLATEX simplifies maintenance. It supports consistent sorting, grouping, and filtering of references, and it allows unused or redundant entries to be identified cleanly. As a project evolves over time, these features help keep the bibliography coherent rather than accreted.

Most importantly, BIBLATEX encourages treating the bibliography as a structured database rather than as formatted text. This perspective aligns well with the workflows described in this chapter: generation, cleaning, completion, and verification of bibliographic data remain distinct steps, each of which can be assisted by ChatGPT while leaving final responsibility with the author.

11.5.1 Examples

To make these capabilities concrete, here are a few things that BIBLATEX can do naturally once bibliographic data are stored in a structured form.

- Produce separate reference lists for different kinds of material. For example, books can be listed separately from journal articles, conference papers, preprints, or online resources. This is particularly useful in long books, where readers may wish to consult monographs differently from research papers.
- Group references by time period. A bibliography can be organized by decades, years, or other temporal categories, making it easy to see how a subject developed over time or to distinguish classical sources from recent work.
- Create multiple bibliographies from the same database. One chapter can list only the references cited there, while another section can present a comprehensive bibliography, without duplicating or reformatting data.
- Present different views of the same bibliography. The same set of entries can be sorted by author, by year, or by title, depending on the purpose, without changing the underlying database.
- Handle modern reference types cleanly. Preprints, datasets, software packages, and online resources can appear alongside traditional books and papers, with appropriate distinctions preserved.
- Support multilingual bibliographies. Titles, notes, and publication information can appear in different languages when appropriate, without forcing everything into a single format or convention.
- Maintain consistency automatically. Once conventions are fixed, names, dates, and titles are handled uniformly across the entire document, reducing the risk of silent drift as the project grows.

These examples illustrate the central idea: with BIBLaTeX, the bibliography is no longer a static list at the end of the document, but a flexible representation of structured data that can be reorganized to suit different mathematical and expository needs.

11.6 Exercises

Exercise 11.1 Ask ChatGPT to generate BibTeX entries for three publications:

1. a book,
2. a journal article,
3. a paper identified by a DOI.

Verify each entry against an authoritative source (MathSciNet, zbMATH, or publisher site).

Identify at least one field that ChatGPT produced incorrectly or incompletely.

Exercise 11.2 Provide ChatGPT with several BibTeX entries that refer to the same work but use different formats (author names, capitalization, keys). Ask it to merge them into a single canonical entry. Verify that no bibliographic information has been lost or altered incorrectly.

Exercise 11.3 Give ChatGPT a small BibTEX database with inconsistent formatting. Ask it to normalize:

1. author names,
2. title capitalization,
3. field ordering.

Compile the document and inspect whether the formatted bibliography reflects the intended style.

Exercise 11.4 Provide an incomplete BibTEX entry (missing publisher, pages, or DOI). Ask ChatGPT to complete it.

1. Identify which fields were added.
2. Verify each added field independently.
3. Decide whether any field should have been left blank rather than guessed.

Exercise 11.5 Ask ChatGPT to scan a BibTEX file for entries with missing required fields.

1. Verify each reported issue manually.
2. Compare the result with BibTEX or biber warnings after compilation.

Exercise 11.6 Give ChatGPT a BibTEX file and a list of citation keys used in your document. Ask it to identify:

1. unused entries,
2. missing citations.

Confirm the results by compiling the document and checking warnings.

Exercise 11.7 Ask ChatGPT to rewrite a set of BibTEX entries to match a specific style (for example, consistent use of initials, journal abbreviations, and punctuation). Inspect the output for unintended changes to bibliographic data.

Exercise 11.8 Provide ChatGPT with a vague bibliographic description (author and year only). Ask it to suggest candidate references and search terms.

1. Use these suggestions to search MathSciNet or ZbMATH.
2. Identify which suggestions were helpful and which were misleading.

Exercise 11.9 Ask ChatGPT to propose a set of keywords and standard terminology for a topic in your research area. Use these terms to perform a literature search. Evaluate whether the suggested terminology improved the quality of the search.

Exercise 11.10 Ask ChatGPT to reorganize a BibTeX database for long-term maintenance:

1. consistent citation keys,
2. normalized fields,
3. minimal redundancy.

Review the result as a database rather than formatted output. Explain which changes improve reliability and which require caution.

11.7 Transition

This chapter has focused on managing bibliographic data: how references are generated, verified, and organized.

In the next chapter, I turn to the index, which serves a complementary role. While the bibliography records sources, the index provides a structured way for the reader to locate definitions, examples, and key ideas throughout the text.

12 Index

I still remember the first index I prepared for one of my books. I expected routine work. I was wrong.

Indexing was slow, tedious, and treacherous. Some entries looked harmless but were not. Others invited duplication, inconsistency, or silent error. Indexing rewards care and punishes haste.

Years later, I made a different mistake. Pressed for time, I trusted an index I had not checked carefully enough. Only after publication did I notice that a central concept appeared under two names, while another important term was missing altogether. The mathematics was correct. The index was not. That experience cured me of taking indexing lightly.

Indexing is still not my favorite task. But I now have a reliable assistant: ChatGPT. It handles much of the mechanical work. I keep control of meaning and structure. That division of labor works.

Before going further, I need to warn you about a subtle trap. The `\index` command looks innocent. It is not.

An index entry is placed where the command is executed, not where the word happens to appear. A definition may span pages. A concept may be motivated first and defined later. If I place `\index` at the wrong spot, the page number will be wrong, even though the source looks correct. Nothing crashes. The index quietly lies.

There is a second deception. The `\index` command has a hidden two-part form. In `\index{Abel@Abel}`, read "abel at Abel", is really "Abel at abel". The part before `@` is the sort key, and the part after `@` is what gets printed.

G. Grätzer, *ChatGPT for Math*, https://doi.org/10.1007/978-3-032-27286-7_12

Principle

An index is a map of the book

Readers use it to find definitions, examples, theorems, notation, and people. ChatGPT can accelerate index preparation, but responsibility for accuracy and coherence remains mine.

12.1 When I Need an Index (and When I Do Not)

I need an index when readers are more likely to search than read straight through. As a rule of thumb, if my work contains many definitions, recurring objects, notation, or named results, it needs an index.

Typical signals are clear:

- the book has several chapters and a substantial table of contents,
- key terms recur across chapters,
- readers are likely to look up notation or symbols,
- many mathematicians are cited by name.

I usually skip an index for a short article, a single lecture note, or a self-contained exposition with little cross-reference.

12.2 The Mechanical Setup in LaTeX

12.2.1 Choosing an Indexing Package

For most of my books, I use `imakeidx`, not the old `makeidx`. It integrates cleanly with my workflow and supports multiple indexes when I need them.

Example

Minimal setup

```
\usepackage{imakeidx}
\makeindex[intoc,title=Index]
```

I place `\printindex` where I want the index to appear, usually at the end of the book.

12.2.2 Compiling the Index

The compilation sequence is simple:

1. run LaTeX to produce the `.idx` file,
2. run the `makeindex` app to produce the `.ind` file (the app is usually listed in the `Typeset` menu),
3. run LaTeX again to typeset the index.

12.3 Designing Indexing Rules Before Inserting Commands

Before inserting a single `\index`, I write down rules. This step matters.
My rules are these:

1. I index only conceptual targets: definitions, core constructions, named objects, notation, and major examples.
2. I choose one canonical form for each entry. Sorting keys are plain text. Printed forms are capitalized and readable.
3. I use subentries to impose structure.
4. I never index in section titles, captions, math mode, or existing `\index` commands.

These rules form the specification I give to ChatGPT.

12.4 Generating Candidate Index Entries

ChatGPT is good at proposing candidates. It identifies structures, notation, examples, and names. It does this quickly.

Example

Candidate list from one section

Concepts: lattice of sublattices; covering relation; atom; coatom.
Notation: N6; ω, β, γ, τ.
People: Birkhoff.

This list is raw material. Some entries will survive. Others will not.

12.5 Collating the Candidate List

Once I have candidates, I normalize them. This is where quality is won.

12.5.1 Merging Variants and Enforcing a Canonical Form

Variants appear everywhere. They must be merged.

Example

Merging variants

join-irreducible element; join irreducibles; join-irreducible elements

I choose one canonical entry and discard the rest.

Example

Symbol normalization table

Variants	Sort → Print	Index command
N6, N_6, N_6	`n6` → N6	`\index{n6@$\mathsfN6$}`
LaTeX, `\LaTeX`	`latex` → LaTeX	`\index{latex@\LaTeX}`
TikZ, tikz	`tikz` → TikZ	`\index{tikz@TikZ}`

Sorting keys are stable. Printed forms are consistent. All other variants are discarded.

12.5.2 Name and Symbol Indexes

Names and symbols deserve special care.

Example

Name index

Abel; `Ábel`; Niels Henrik Abel.

I choose one form and write

```
\index{abel@Abel, Niels Henrik}
```

In the index, this appears as

Abel, Niels Henrik 23, 47

Example

Subentries and their printed form

```
\index{join-irreducible element@Join-irreducible element}
\index{join-irreducible element!definition@Definition}
\index{join-irreducible element!examples@Examples}
```

Printed index:

Join-irreducible element 12
    Definition 13
    Examples 18, 21

12.6 ChatGPT Inserts Index Commands

After normalization, I let ChatGPT insert `\index` commands. Under strict rules!

Example

Insertion report

When I ask ChatGPT to insert `\index` commands, I require two outputs.

Edited LaTeX

ChatGPT returns the full edited section with the `\index{...}` commands inserted.

(continued)

Insertion report

Immediately after the edited LaTeX, ChatGPT prints a list of insertions. Each line contains (i) the exact `\index{...}` inserted and (ii) the sentence where it was inserted.

▶ Warning
Uncontrolled automation ruins indexes

Without constraints, ChatGPT will make mistakes that break the document.

12.7 Final Review

Before shipping the manuscript, I check the index by hand, verify meaning, placement, and structure.

Principle

Trust but verify

ChatGPT is fast. The responsibility remains mine.

Indexing resembles drawing diagrams and checking proofs. The tools are powerful. The speed is intoxicating. But correctness remains a human obligation.

12.8 Exercises

Exercise 12.1 Provide ChatGPT with a BibTeX file containing at least ten entries. Ask it to identify entries with missing required fields (for example, `year` in `@article` or `publisher` in `@book`). Verify each report against the BibTeX specification.

Exercise 12.2 Give ChatGPT several BibTeX entries with author names in different formats (initials, full names, particles such as `von`, accented characters). Ask it to normalize the author fields while preserving correct name parsing. Inspect the result by running BibTeX and checking the formatted bibliography.

Exercise 12.3 Provide two BibTeX entries that refer to the same work but use different keys. Ask ChatGPT to construct a single canonical entry and to list which fields were merged, discarded, or rewritten. Confirm that no bibliographic information was lost.

Exercise 12.4 Ask ChatGPT to scan a BibTEX file for inconsistent capitalization in titles and to correct it using protected braces where appropriate. Check that mathematical symbols and proper nouns remain unchanged in the compiled bibliography.

Exercise 12.5 Introduce deliberate syntax errors into a BibTEX entry (unmatched braces, missing commas, unescaped special characters). Ask ChatGPT to locate and repair the errors. Verify the fix by compiling the document without BibTEX warnings.

Exercise 12.6 Give ChatGPT a BibTEX file and a list of citation keys used in the text. Ask it to identify unused entries and missing citations. Confirm the result using BibTEX or `biber` diagnostics.

Exercise 12.7 Ask ChatGPT to convert a small BibTEX database from one entry type to another (for example, `@misc` to `@online` or `@inproceedings` to `@article`). Check that the field mapping is appropriate and standards-compliant.

Exercise 12.8 Provide ChatGPT with a BibTEX file containing URLs and DOIs. Ask it to verify field placement and formatting conventions (`doi` vs. `url`, use of `howpublished`, etc.). Confirm consistency with the bibliography style you are using.

Exercise 12.9 Ask ChatGPT to reorder fields within BibTEX entries according to a fixed canonical order. Verify that the reordering does not affect compilation or output.

Exercise 12.10 Give ChatGPT a BibTEX file and ask it to prepare a version suitable for long-term maintenance: consistent keys, normalized fields, and minimal redundancy. Review the result as a database, not as formatted output.

12.9 Transition

The index provides access to what is already in the book, but its usefulness depends on consistent naming.

In the next chapter, I address notation and terminological consistency, ensuring that the same ideas are referred to in stable and predictable ways throughout the text.

Notational and Terminological Consistency 13

One of the quiet challenges of mathematical writing is maintaining consistent notation and terminology. A paper or book evolves over months or years. Over time, new symbols are introduced, older ones are redefined, terminology is adopted from multiple sources, and material from earlier drafts is merged. As a result, notation drifts. ChatGPT is remarkably effective at detecting these inconsistencies and proposing coherent, uniform alternatives.

13.1 Why Consistency Matters

Consistent notation helps readers follow a long argument without stumbling. Inconsistent notation forces readers to remember unnecessary details: whether f^* or f' denotes a certain morphism, whether $J(x)$ means the ideal generated by x in one chapter and something else in another, or whether "covering relation" means the strict covering relation rather than a relation that includes the equality case.

ChatGPT cannot enforce consistency automatically, but it can perform several supporting tasks.

- detect conflicts,
- propose unifying conventions,
- rewrite sections with consistent symbols,
- generate glossaries of notation and terminology.

13.2 Detecting Inconsistent Notation

Inconsistent notation is easy to create. It is hard to see. By the time I prepare the index, it is already present.

G. Grätzer, *ChatGPT for Math*, https://doi.org/10.1007/978-3-032-27286-7_13

The problem is cumulative. A symbol is introduced early. Later, a variant slips in. Still later, a third form appears. The source may be ChatGPT, another author, or an earlier draft. Each occurrence looks reasonable in isolation. Together, they are a mess.

Readers notice this immediately. They search for one symbol and encounter several. They are forced to guess. An index that reflects this inconsistency only makes the problem worse.

This is one place where ChatGPT helps me. I ask it to scan a chapter and list all symbols and technical terms. What matters is not correctness but completeness. I want to see everything.

Once I have that list, patterns emerge. Variants cluster. Near-duplicates reveal themselves. I then decide what survives.

Here the standard workflow introduced in the Introduction is used to detect, test, and repair global inconsistencies.

Example

Detecting inconsistent notation

From one chapter, ChatGPT reports the following symbols and phrases:

N6, N_6, six-element lattice

Each instance is fine. Taken together, they signal inconsistency.

I choose one notation, N6, and treat it as canonical. All other forms are rewritten in the text or redirected in the index using a stable sorting key.

The inconsistency disappears. The index becomes coherent.

13.3 Creating Glossaries

An index tells readers where things appear. A glossary tells them what those things mean. They serve different purposes.

I create a glossary when terminology is dense, when definitions are brief, or when readers may consult the book nonsequentially. This happens often in technical writing, especially in books that mix text, symbols, and examples.

A glossary is selective. I do not include every defined term. I include terms that readers are likely to forget, confuse, or want to check quickly. Short definitions work best.

ChatGPT is helpful here, but in a different role than for indexing. I do not ask it to invent definitions. I ask it to extract and standardize them. The authority remains with me.

My workflow is simple. I identify glossary candidates, decide which ones belong, and then normalize wording and notation. Only after that do I format the glossary.

Example

Glossary extraction and normalization

From one chapter, I ask ChatGPT to extract glossary candidates. It returns the following list:

atom, coatom, covering relation, lattice of sublattices

I decide which entries belong. I rewrite definitions for uniform tone and length. For example, I finalize:

Atom. An element that covers the least element of a lattice.
Coatom. An element covered by the greatest element of a lattice.
Covering relation. The relation $x \prec y$ indicating that $x < y$ and no element lies strictly between them.

The glossary is now consistent, concise, and independent of page numbers.

13.4 Rewriting for Uniform Style

Once notation and terminology are unified, ChatGPT can rewrite a section for uniform stylistic tone. For example:

Example

Prompt

Rewrite this subsection in a consistent style using formal mathematical prose, avoiding conversational expressions, and keeping the notation fixed.

Or:

Example

Prompt

Rewrite this chapter in a slightly more elementary style, keeping all definitions and theorems while simplifying transitions and notation.

This is particularly helpful when combining material originally written in different contexts.

13.5 Detecting Unused or Redundant Symbols

Over time, some notation becomes obsolete. Symbols may be defined early and never used. Variables may be introduced in a lemma and fail to appear later. ChatGPT can identify such cases.

> List all symbols defined in this section that are never used afterward.

The model returns a short list of extraneous notation that can be removed or simplified.

13.6 Global Consistency Checks

Local consistency is not enough. A book is a unified entity, not just a collection of parts. Notation, terminology, and indexing decisions must agree across chapters.

This is where small inconsistencies hide. A symbol is corrected in one chapter but not in another. A term is capitalized in the index but appears in lowercase elsewhere. Individually, these slips are minor. Taken together, they erode trust.

At this stage, I ask ChatGPT to scan the entire manuscript. I do not ask it to fix anything. I ask it to report. The output is a checklist for me, not a verdict.

ChatGPT produces a report of this form:

Example

Global consistency report

> Symbol variants detected: N6, N_6, N_6 (Chapters 2, 5, 7).
> Terminology variants: "join irreducibles" versus "join-irreducible elements" (Chapters 3, 6).
> Index capitalization mismatch: "Lattice of sublattices" versus "lattice of sublattices".
> Name variants: Abel, N. H. versus Abel, Niels Henrik (Chapters 1, 4).
> Glossary overlap: "atom" defined twice with different wording (Chapters 2, 8).

Each line points to a decision I must make. Once these are resolved, the manuscript becomes internally coherent.

13.7 Exercises

Exercise 13.1 Give ChatGPT a BibTEX file used across multiple projects. Ask it to propose a stable key-naming scheme that minimizes collisions and remains readable over time. Check the proposal against your existing citation keys.

Exercise 13.2 Provide BibTEX entries containing mathematical notation in titles. Ask ChatGPT to insert protective braces correctly so that capitalization and symbols are preserved by the bibliography style. Verify the result by compiling the document.

Exercise 13.3 Ask ChatGPT to scan a BibTEX file for fields that are style-dependent (for example, `number`, `month`, or `note`) and to flag entries where these fields may cause inconsistent output. Confirm the behavior by switching bibliography styles.

Exercise 13.4 Give ChatGPT a BibTEX database that mixes legacy and recently added entries. Ask it to identify outdated conventions (for example, obsolete entry types or deprecated fields) and to modernize them while preserving content. Review the changes entry by entry.

Exercise 13.5 Provide a BibTEX file and ask ChatGPT to extract a minimal subset containing only entries cited in a given chapter. Compile the chapter using the reduced file and confirm that no references are missing.

Exercise 13.6 Ask ChatGPT to analyze a BibTEX file for portability: character encoding, use of macros, and dependence on external files. Determine which features may fail when the project is moved to a different system or publisher.

Exercise 13.7 Introduce subtle semantic errors into BibTEX entries (for example, incorrect entry types or misplaced fields). Ask ChatGPT to detect and correct them. Verify that the corrected entries produce the intended output.

Exercise 13.8 Ask ChatGPT to prepare documentation for a BibTEX database: a short description of key conventions, field usage, and maintenance rules. Evaluate whether the documentation would be sufficient for a future collaborator.

13.8 Transition

With notation and terminology made consistent, we return to mathematical work at a broader level.

In the next chapter, I examine how ChatGPT can assist in exploratory mathematical research, including reformulating problems, testing conjectures, and identifying connections to known results.

AI-Assisted Research 14

Mathematical research is not a single activity. It involves exploring small cases, reformulating definitions, testing hypotheses, constructing examples, searching the literature, and planning proofs. ChatGPT cannot replace mathematical reasoning, but it excels at supporting the exploratory phase of research. In this chapter, I describe practical ways to use ChatGPT as a research assistant—one that helps us view a problem from different angles, test ideas quickly, and organize our thinking.

14.1 Reformulating a Problem

When ChatGPT fails, the problem is often mine. I asked the wrong question.

Mathematical problems are rarely stated in the form that makes them easiest to solve. They are stated in the form that reflects how I think. ChatGPT reflects that wording back to me. If the wording is poor, the response is poor.

Reformulation is not cheating. It is a standard mathematical move. I do it when I change notation, I pass to an equivalent definition, or when I replace a vague question by a precise one.

With ChatGPT, reformulation is especially powerful. I can ask the same question several times. I can vary the wording slightly from attempt to attempt. I then compare the responses. Patterns emerge quickly.

The goal is not elegance, it is traction.

Reformulating a problem, my first prompt was:

Example

Prompt
Draw the lattice of sublattices of N6.

G. Grätzer, *ChatGPT for Math*, https://doi.org/10.1007/978-3-032-27286-7_14

The result was unusable. Too many edges. No structure. No labels.

I reformulated the problem:

Example

Prompt

Draw the lattice of sublattices of N6. Place circles at distinct levels by cardinality. Use blue edges behind purple nodes. Rotate the diagram sideways. Label each node with a letter.

The mathematics did not change. The answer did. See Fig. 14.1.

14.2 Identifying Minimal Hypotheses

When a statement is true, it is often true for reasons simpler than the ones I first give. I add assumptions because they feel safe. They may be unnecessary.

Identifying minimal hypotheses is a standard mathematical task. ChatGPT is useful here because it has no emotional attachment to my assumptions. It will happily test weaker ones. It will remove structure. It will propose counterexamples.

I do not ask it to prove the strongest version. I ask it to weaken the hypotheses.

The answer is rarely final. But it is informative.

Example

Identifying minimal hypotheses

I began with the following claim:

In a finite, distributive lattice, every congruence is determined by its values on the join-irreducible elements.

I asked ChatGPT:

Prompt

Which hypotheses in this statement are actually used? Can any of them be weakened?

ChatGPT suggested testing semidistributivity and dropping finiteness. This immediately forced me to think. Where did finiteness enter? Was distributivity essential or only join-semidistributivity?

The original statement did not change. My understanding did.

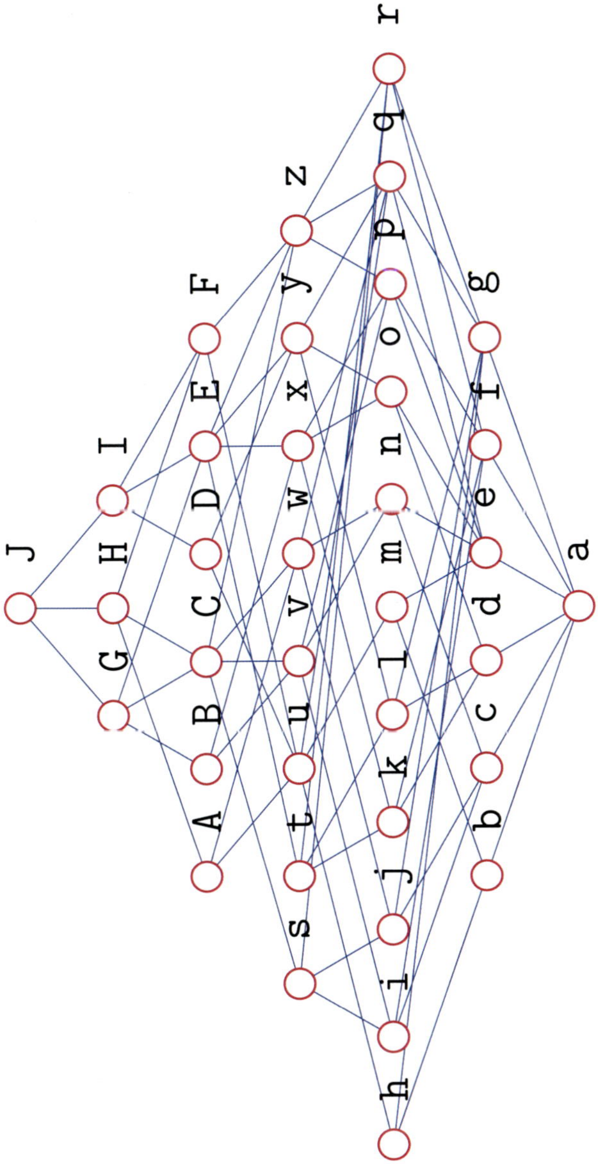

Fig. 14.1 The lattice of all sublattices of N6, drawn sideways

14.3 Small Cases and Diagrams

When structure matters, I draw. Small cases become informative only when I can see them.

ChatGPT is useful here because it generates diagrams quickly. I can vary the example, inspect the result, decide what is structural and what is incidental.

Example

Diagram-based exploration

I explored three minimal lattices. They were a chain, a square, and N_6.

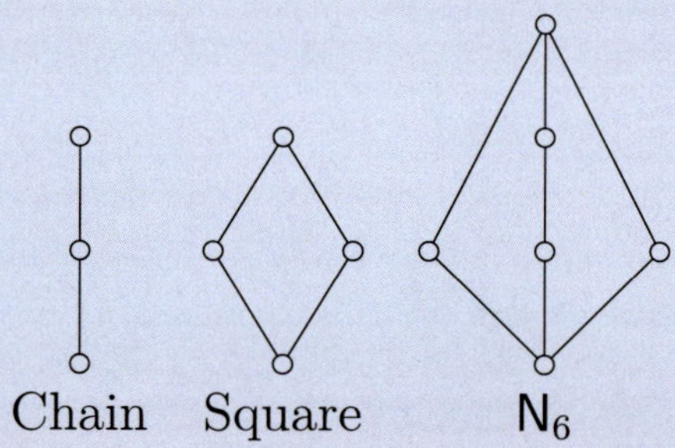

Looking at these diagrams, the differences are immediate. In a chain, congruences are forced. In the square, symmetry constrains them. In $\mathsf{N}+6$, a single additional relation changes the behavior entirely.

No proofs. But the pictures told me where a proof might exist and where it could not.

14.4 Constructing Examples

When I suspect a statement is false, I try to break it. When I suspect it is true, I try to stress it. In both cases, I look for small lattices.

Diagrams are ideal for this work. A diagram can function as an example. It can also function as a counterexample. The difference is not the picture. It is the property I test.

ChatGPT helps me by generating candidate diagrams quickly. I then verify the structure myself, check the cover relations, and check joins and meets. Only then do I trust the example.

Example

Examples and counterexamples by diagrams

To test a claim, I assemble a small gallery of lattices. Even three diagrams can be decisive.

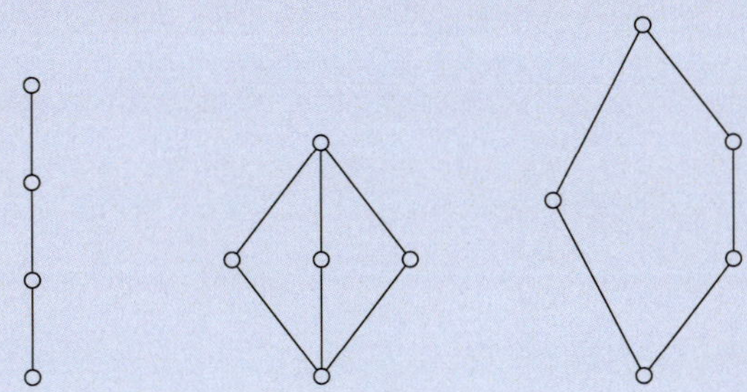

Chain (C4) Diamond (M3) Pentagon (N5)

Suppose that a claim is asserted for all finite lattices. I first test it on C4. If it fails there, I am done. If it survives, I test M3 and N5. Many "obvious" statements collapse immediately on M3 or N5.

These diagrams do not replace a proof. They tell me what is plausible. They often point to a missing hypothesis.

14.5 Conjectures

14.5.1 Generating Conjectures

One of the most natural uses of ChatGPT in mathematics is to help *generate conjectures*. Given data, examples, or a small class of structures, ChatGPT can often detect patterns and suggest plausible general statements. Of course, these suggestions are not proofs; they are starting points for mathematical work.

The typical workflow is:

- supply a family of examples or computed data,
- ask ChatGPT to look for regularities,
- extract a conjecture that can then be tested or proved.

Example: congruence lattices of small lattices.

Example

Prompt

For each of the following lattices, compute the number of congruences:

$$\mathsf{C}_2, \quad \mathsf{C}_3, \quad \mathsf{M}_3, \quad \mathsf{N}_5$$

Look for a pattern and suggest a conjecture relating the number of congruences to structural properties of the lattice.

ChatGPT response (abridged).

- The two-element chain C_2 has 2 congruences.
- The three-element chain C_3 has 4 congruences.
- The lattice M_3 has 2 congruences.
- The lattice N_5 has 5 congruences.

Chains seem to have many congruences, while nondistributive lattices have comparatively few. This suggests that distributivity increases the number of congruences.

Extracted conjecture.

Example

Among finite lattices of a fixed size, distributive lattices tend to maximize the number of congruences, while nondistributive lattices tend to have fewer congruences.

Discussion. The conjecture is imprecise as stated but it points in a meaningful direction. An expert will immediately recognize connections with the fact that finite distributive lattices have Boolean congruence lattices, while M_3 only has few. The role of ChatGPT here is not to replace theory, but to *propose a hypothesis* that guides further investigation.

In practice, I often refine such conjectures by asking ChatGPT to:

- test additional small cases,
- restate the conjecture more sharply,
- suggest known results that might confirm it or suggest results that might refute it.

14.5.2 Testing Conjectures

Once a conjecture has been formulated, the next step is to test it against additional evidence. ChatGPT is particularly useful at this stage: it can rapidly examine small cases, look for counterexamples, and clarify the limits of a proposed statement. A single counterexample is often more valuable than many confirming instances. The goal at this stage is not proof, but *stress testing*:

- checking boundary cases,
- exploring nearby structures,
- and attempting to disprove the conjecture.

Example

Conjecture

Among finite groups with a fixed number of elements, abelian groups have more normal subgroups than nonabelian ones.

Prompt

Test the conjecture for all groups with at most 12 elements. List the number of normal subgroups for each group type and indicate whether the conjecture holds in each case. If it fails, exhibit a counterexample.

ChatGPT response (abridged)

- Cyclic groups of order n have one normal subgroup for each divisor of n.
- The Klein four-group has 5 normal subgroups.
- The symmetric group S_3 has 3 normal subgroups.
- The dihedral group D_4 has fewer normal subgroups than the abelian groups of the same order.

In these cases, the abelian groups have strictly more normal subgroups than the nonabelian ones. No counterexample appears among groups with at most 12 elements.

Assessment. The conjecture survives testing in small cases, but remains unproven. At this point, the conjecture can be strengthened. It can be weakened to avoid hidden assumptions. It can also be targeted toward a proof strategy.

This illustrates a typical use of ChatGPT in conjecture testing: it acts as a fast, tireless assistant that checks cases an expert would not want to enumerate by hand, while leaving judgment and proof to the mathematician.

14.5.3 Connecting the Problem to Known Results

After a conjecture has survived initial testing, a natural next step is to connect it with existing theory. ChatGPT can be helpful here by suggesting relevant theorems, pointing to classical results, and proposing well-known frameworks into which the conjecture might fit. This often turns an informal observation into a mathematically grounded question.

At this stage, the goal is to move from experimentation to context. One identifies related theorems or classifications. One recognizes standard constructions or invariants and places the conjecture within the existing literature.

14.6 Strategies for Proofs

After a conjecture has been formulated, tested, and connected to known results, the next step is to plan a proof. ChatGPT can be useful at this stage by suggesting possible strategies. It can identify relevant tools. It can outline plausible lines of argument. These outlines do not replace a proof, but they often help structure the work.

At this stage, the goal is to move from statement to method:

- decide which known results might be invoked,
- identify key structural features to exploit,
- and break the proof into manageable steps.

14.7 Exploring Analogies

Another productive way to think about a problem is to explore analogies with better-understood situations. ChatGPT is often effective at suggesting such analogies. It draws connections between different areas of mathematics and proposes parallel statements that may illuminate the original problem.

The goal here is not the transfer of results, but the transfer of insight. I identifiy a familiar setting with similar structure, and compare definitions and invariants. Then I can see which arguments might carry over.

Example

Prompt

> Find an analogy between congruences on lattices and familiar equivalence relations in algebra. Explain how this analogy might help in understanding congruence counts.

(continued)

ChatGPT response (abridged)

Congruences on a lattice play a role similar to normal subgroups in group theory. Just as normal subgroups correspond to homomorphic images of a group, congruences correspond to quotient lattices. Groups with many normal subgroups tend to have many quotients, while simple groups have only two. Similarly, distributive lattices tend to have many congruences, whereas lattices such as M3 and N5 have few.

Analogy

Viewing congruences as lattice-theoretic analogues of normal subgroups provides intuition for why congruence counts vary so widely. In both settings, structural rigidity limits the number of admissible quotients, while decomposability increases it.

Usefulness

The analogy does not yield a proof, but it suggests where to look: classification results, minimal obstructions, and quotient structures. In practice, such analogies often guide the search for the right definitions and theorems.

14.7.1 A Warning: The Role of Verification

Exploration is useful, but verification remains essential (see Chapter 3). ChatGPT may propose patterns that do not extend beyond the small cases, or proof strategies that ignore critical hypotheses. It is our responsibility to verify the details, test every claim, and refine every strategy.

14.8 Exercises

Exercise 14.1 Take a research problem you are currently considering. Ask ChatGPT to restate the problem at three levels: informal intuition, precise high-level formulation, and fully formal mathematical statement. Compare the three versions and decide which best clarifies the core issue.

Exercise 14.2 Ask ChatGPT to list known results or standard techniques that are plausibly related to your problem. Verify each suggested connection independently and discard those that rely on nonexistent or irrelevant results.

Exercise 14.3 Choose a conjecture that you believe to be true. Ask ChatGPT to test it on small or special cases. Check each reported case by hand or by computation.

Exercise 14.4 Ask ChatGPT to suggest alternative approaches to your problem (for example, algebraic, combinatorial, or order-theoretic). Identify which suggestions correspond to established methods and which are superficial analogies.

Exercise 14.5 Provide ChatGPT with a partial proof or proof outline. Ask it to identify points where assumptions might be weakened or where a different lemma could simplify the argument. Evaluate these suggestions critically.

Exercise 14.6 Ask ChatGPT to generate a plausible but incorrect argument related to your problem. Analyze the argument carefully and locate the precise point of failure.

14.9 Transition

This chapter has focused on exploration in mathematical research, where ideas, conjectures, and strategies are tested rather than trusted.

In the next chapter, I turn to explanation: how mathematical ideas can be communicated clearly to different audiences, and how ChatGPT can assist in this process without compromising precision.

Explaining Math 15

Mathematics is not only about proving theorems; it is also about *explaining* them. I explain to myself as I try to understand a new concept, to colleagues who share my background but not my notation, and to students who may be seeing an idea for the first time. In all these situations I adapt the level of detail, the choice of examples, and the vocabulary to my audience.

ChatGPT is particularly well suited to this explanatory side of math. It works in language, and it is comfortable shifting between levels: from informal to formal, from undergraduate to researcher, from algebraic to geometric. In this chapter I explore how I, as a working mathematician, can use ChatGPT as an assistant for explanation. It can produce multi-level expositions, translate between different "dialects" of math, rewrite technical passages for different audiences, and suggest analogies and metaphors.

For a systematic approach to verifying generated mathematical text, see Chapter 3.

15.1 The Challenge of Mathematical Explanation

Explaining math is harder than it looks. A proof that is logically correct may be nearly unreadable. A definition that is perfectly precise may remain opaque to a student who has no picture of what it means. Conversely, an informal explanation that is vivid may hide important hypotheses or gloss over subtle points.

When I explain mathematics, I face at least three simultaneous constraints:

- *Precision.* I must respect logical structure: the quantifiers, the hypotheses, the dependence of each step on the previous ones.
- *Clarity.* The reader must not be forced to carry too many unresolved questions at once. The flow of ideas should be visible, not buried under notation.
- *Appropriateness.* The level of abstraction, the amount of background assumed, and the kind of examples chosen must match the audience.

G. Grätzer, *ChatGPT for Math*, https://doi.org/10.1007/978-3-032-27286-7_15

15.2 Multi-level Explanations of a Single Concept

One of the simplest and most effective uses of ChatGPT is to ask for several explanations of the *same* concept, each tailored to a different audience.

15.2.1 A Basic Workflow

Suppose I want to explain a fundamental distributive property in linear algebra. I proceed in three steps.

Example

Prompt

Explain distributivity in linear algebra at three levels

Definition

Let V be a vector space over a field $\mathbb{F}$. Vector addition and scalar multiplication satisfy the distributive law: for all vectors $u, v \in V$ and all scalars $\lambda \in \mathbb{F}$,

$$\lambda(u + v) = \lambda u + \lambda v.$$

(a) Strong undergraduate

This law says that scaling a sum of two vectors gives the same result as scaling each vector separately and then adding the results. If you picture vectors geometrically, stretching the vector $u + v$ by a factor λ gives the same vector as stretching u and v individually and then adding them. Because of this property, computations with vectors behave in a familiar, predictable way, much like arithmetic with numbers.

(b) Beginning graduate student in linear algebra

Distributivity expresses the compatibility between the additive and scalar structures on a vector space. It ensures that scalar multiplication is a linear operation with respect to vector addition. This property is essential for the theory of linear maps, since it guarantees that linear transformations preserve addition and can be analyzed using algebraic identities. Without distributivity, the basic framework of linear algebra would collapse.

(c) Expert in another area of mathematics

From a structural viewpoint, the distributive law encodes the interaction between two algebraic operations defining a module over a field. It ensures

(continued)

that scalar multiplication defines an additive endomorphism of the underlying abelian group. This compatibility underlies the categorical interpretation of vector spaces as modules and supports the use of linear algebraic methods across analysis, geometry, and applied mathematics.

Typically, the undergraduate explanation relies on concrete vector calculations and geometric intuition. The graduate-level explanation emphasizes algebraic structure and linearity. The "expert outsider" version highlights how the distributive law situates vector spaces within a broader mathematical framework.

15.2.2 Benefits of the Multi-level Approach

This simple procedure has several benefits for me.

- It forces me to separate the *core idea* from the *technical machinery*. Anything that appears in all three explanations is likely to be central; material that appears only in the expert version is probably background or context.
- It reveals missing intuition. If the undergraduate version feels vacuous, that is a sign that I need a better example or a more visual description.
- It provides ready-made material for lectures, office hours, and written notes. I can keep a short file of multi-level explanations for recurrent concepts in my courses.

In practice, the explanations ChatGPT produces are rarely perfect. They may be slightly too verbose, or they may use terminology I would not choose. The point is not to accept them uncritically, but to use them as raw material for my own refined version.

15.3 Translating Between "Dialects"

Most mathematical ideas can be expressed in more than one language. A statement about lattices can be rewritten as a statement about closure operators. A topological condition can be reformulated in terms of nets or filters. Many categorical notions have concrete incarnations in familiar algebraic settings. Skilled mathematicians move between these dialects as needed, but the translation is not always obvious.

ChatGPT can act as my assistant translator.

15.3.1 From One Subfield to Another

I might ask:

Example

Prompt

Explain a group-theoretic result to a module expert

Group-theoretic statement

Let

$$0 \longrightarrow A \longrightarrow B \longrightarrow C \longrightarrow 0$$

be a short exact sequence of abelian groups. If C is free abelian, then the sequence splits.

Explanation for someone who thinks in modules

First, reinterpret the statement categorically. An abelian group is nothing more than a $\mathbb{Z}$-module. So the problem is not really about groups at all; it is about modules over the ring $\mathbb{Z}$.

In module language, the statement becomes: every short exact sequence of $\mathbb{Z}$-modules splits if the quotient module C is free.

Now recall the module-theoretic fact you already know: free modules are projective. Projective modules are precisely those modules P for which every surjection onto P admits a section. That is exactly the definition of a splitting short exact sequence.

So the group-theoretic conclusion follows immediately: since C is free as a $\mathbb{Z}$-module, it is projective, hence the surjection $B \to C$ has a section, and therefore $B \cong A \oplus C$.

What looks nontrivial in group language

If one insists on thinking purely in terms of groups, the proof appears to require explicit constructions using generators, lifting basis elements, and checking homomorphism properties. From the module viewpoint, all of this collapses into a single statement: "free implies projective".

Moral

Many structural results about abelian groups are not really group-theoretic phenomena. They are consequences of basic module theory over $\mathbb{Z}$. Once the translation is made, the result becomes automatic.

15.3.2 Aligning Notation and Terminology

Even within a single field, notation and terminology vary. One paper may write $\leq$ where another writes $\subseteq$; one book may say "ideal", another "lower set". ChatGPT can help me harmonize these choices when I prepare a survey or a set of lecture notes.

Example

Prompt
Rewrite in the style of Birkhoff

Original text

> A lattice is a structure that allows us to combine information in two complementary ways. The join operation represents aggregation or choice, while the meet operation captures common constraints. From this point of view, lattices generalize Boolean logic and appear naturally in computer science, data analysis, and optimization.
> In a distributive lattice, these two operations interact cleanly: computing a meet after forming a join gives the same result as combining meets first. This property ensures that no hidden dependencies arise when elements are merged in stages, making distributive lattices especially well behaved.

Rewritten version (Birkhoff-style)

> A lattice is an ordered set in which every pair of elements has a least upper bound and a greatest lower bound, denoted by $\vee$ and $\wedge$, respectively. These operations determine the order structure and suffice for the algebraic description of the lattice.
> A distributive lattice is one in which the operations $\wedge$ and $\vee$ satisfy the distributive identities
>
> $$x \wedge (y \vee z) = (x \wedge y) \vee (x \wedge z),$$
> $$x \vee (y \wedge z) = (x \vee y) \wedge (x \vee z).$$
>
> In such lattices, the formation of joins and meets is compatible in the sense that either operation may be applied first without affecting the result. This condition excludes certain finite configurations and characterizes those lattices whose order-theoretic behavior is closest to that of the lattice of subsets of a set.

15.4 ChatGPT as an Informal Teaching Assistant

Although this book focuses on the mathematician's perspective, many of the explanatory workflows I have described can also be used by students, under appropriate guidance.

15.4.1 Office-Hours Style Conversations

Students can use ChatGPT as a simulated office hour in a very literal way. A student may paste a definition or theorem directly from the course notes into ChatGPT and ask for a step-by-step explanation of the statement and its hypotheses. They may then ask for explicit examples, counterexamples, or small diagrams illustrating the concept. This does not replace interaction with a human instructor, but it can remove the initial barrier of not knowing how to begin.

As an instructor, I can anticipate this behavior and guide it by providing concrete, well-formed prompts that model productive questions. For example, I can suggest prompts such as:

- "Rewrite this definition in plain language, then give two explicit examples and one explicit non-example, explaining each choice."
- "Apply this theorem to the lattice shown in Figure X. Verify each hypothesis directly from the diagram and state whether the conclusion holds."
- "Identify which hypotheses of the theorem fail in this example, and explain precisely where the proof breaks down as a result."

15.4.2 Generating Practice Questions

ChatGPT can also be used to generate short, conceptual self-test questions. For example, I may ask it to produce a small set of questions that check whether a student has understood a definition at the level of ideas, not proofs.

Example

Generating self-test questions

Prompt

> Write ten short questions that test basic understanding of the definition of an abelian group. Each question should be answerable in one sentence. Provide a brief model answer for each question.

Candidate

- What is an abelian group?
  An abelian group is a group whose operation is commutative.
- Which properties must a set and operation satisfy to form a group?
  They must satisfy closure, associativity, have an identity element, and give each element an inverse.

Example

- *Is every group abelian?*
  No, only those groups whose operation is commutative are abelian.
- ....

I then curate the generated questions. For example, I may discard questions that presuppose material not yet covered and edit the model answers to reflect exactly the level of precision I expect from students. The resulting list can be used unchanged as a self-test or lightly edited for an in-class exercise.

15.5 Limitations and Best Practices

As always, the use of ChatGPT comes with limitations.

15.5.1 Checking for Accuracy

This is really important. DO NOT SKIP IT!

Example

Slight inaccuracies

The following are examples of *slight inaccuracies* of the kind that often appear in informal explanations or AI-generated text. Each statement sounds plausible and is close to the truth, but is mathematically incorrect or misleading in a subtle way.

1. **Missing hypothesis**

   Inaccurate statement:
   Every vector space has a basis.

   Issue: The underlying field is not specified, and choice is silently assumed.
   Correction: Every vector space over a field has a basis; this statement generally relies on the Axiom of Choice.

(continued)

2. **Incorrect implication**

   Inaccurate statement:
   Every linearly independent set spans the vector space.

   Issue: Linear independence alone does not imply spanning.
   Correction: A set that is both linearly independent and spanning is a basis.

3. **Overuse of informal language**

   Inaccurate explanation:
   Linear independence means that none of the vectors depends on the others.

   Issue: The defining condition involving linear combinations is suppressed.
   Correction: A set of vectors is linearly independent if the only linear combination equal to the zero vector is the trivial one.

4. **Incomplete characterization**

   Inaccurate statement:
   A vector space is finite-dimensional if it has a spanning set.

   Issue: Infinite-dimensional spaces also have spanning sets.
   Correction: A vector space is finite-dimensional if it has a *finite* spanning set.

5. **Confusion between equality and isomorphism**

   Inaccurate statement:
   Every n-dimensional vector space is $\mathbb{F}^n$.

   Issue: Structural equivalence is mistaken for equality.
   Correction: Every n-dimensional vector space over $\mathbb{F}$ is *isomorphic* to $\mathbb{F}^n$.

6. **Vague dimension reasoning**

   Inaccurate claim:
   Adding one vector increases the dimension by one.

   Issue: Linear dependence is ignored.
   Correction: Adding a vector increases the dimension only if the vector is not in the span of the existing set.

7. **Imprecise definition**

   Inaccurate definition:
   A subspace is any subset closed under addition.

(continued)

Issue: Closure under scalar multiplication and containment of the zero vector are omitted.
Correction: A subspace is a subset containing the zero vector and closed under addition and scalar multiplication.

8. **Unjustified proof shortcut**

   Inaccurate phrase:
   It is obvious that the map is linear.

   Issue: The defining properties of linearity are not verified.
   Correction: We verify explicitly that $f(x+y) = f(x)+f(y)$ and $f(\lambda x) = \lambda f(x)$ for all vectors x, y and scalars λ.

15.6 Patterns and Prompts

15.6.1 Translation Between Dialects

Rewrite this statement using the terminology and notation of [author/book], and highlight any differences in hypotheses or conventions.

15.6.2 Rewriting for a New Audience

Rewrite this section as if it were a set of lecture notes for a first graduate course, adding informal explanations and at least two simple examples.

15.7 Exercises

Exercise 15.1 Choose a mathematical concept you know well. Ask ChatGPT to explain it at three levels: for a general reader, for an advanced undergraduate, and for a beginning graduate student. Check each explanation for accuracy and appropriateness of level.

Exercise 15.2 Take a paragraph of mathematical exposition from your own writing. Ask ChatGPT to rewrite it for greater clarity without changing the mathematical content. Compare the revised version with the original and decide which changes you would accept.

Exercise 15.3 Ask ChatGPT to explain the same idea using two different metaphors or informal analogies. Evaluate which analogy clarifies the concept and which risks misleading the reader.

Exercise 15.4 Provide ChatGPT with a definition and ask it to list common points of confusion or misinterpretation. Decide which of these are genuine risks and how you would address them in exposition.

Exercise 15.5 Ask ChatGPT to produce an explanation that is intentionally too informal and another that is overly formal. Analyze what is gained and lost in each case.

Exercise 15.6 Give ChatGPT a short proof and ask it to rewrite the argument as a narrative explanation, without formulas. Check that the logical structure of the proof is still recognizable.

15.8 Transition

This chapter has focused on explaining mathematics clearly and appropriately for different audiences.

In the next chapter, I turn to the technical realization of such explanations: the use of LaTeX, diagrams, and writing tools to produce clear and reliable mathematical documents in practice.

LaTeX, Diagrams, and Writing 16

For most mathematicians, LaTeX is as essential as the pencil or the blackboard. I use it to

1. write articles and books,
2. prepare lecture notes and slides,
3. draw diagrams.

Over the years I have accumulated a personal ecosystem of macros, style files, shell scripts, and small tricks for converting images and managing bibliographies. This chapter examines how ChatGPT can help with this "mechanical" side of math.

In earlier chapters I have already used ChatGPT to assist with diagrams. I have also used it for interactive editing of mathematical prose and for decoding LaTeX log files. Here I systematize these uses. I also add new ones. I use ChatGPT to

- convert messy text into clean LaTeX,
- document and debug style files,
- do small automations that save time without changing the mathematics.

I also return to indexes and glossaries. My goal is simple. I want less friction. I want more of my time to go to the mathematics rather than to the tools.

16.1 From Messy Text to Structured LaTeX

A common problem in my everyday mathematical life is that material arrives in the wrong format. A collaborator may send an email containing a proof. A publisher may forward an old Word document. A scanner may produce OCR with broken formulas and strange encoding. Before any of this can be used in a book or paper, it must be converted into clean LaTeX.

G. Grätzer, *ChatGPT for Math*, https://doi.org/10.1007/978-3-032-27286-7_16

16.1.1 Basic Conversion Workflow

ChatGPT is surprisingly good at this kind of conversion. A basic workflow is:

1. Paste the raw text, marking clearly where it begins and ends.
2. Specify the desired format:

> Convert this text into LaTeX, using display math for longer formulas. Create theorem environments where appropriate. Use lemma and proof environments when the structure calls for them.

3. Ask for the result to be given as a single `verbatim` block or inside a `lstlisting` environment so that it can be copied safely.

Here are three typical examples.

Example

1. Plain text to theorem and proof

Prompt

Prove:

> Let n be an integer. If n is even, then n^2 is even. Proof. If n is even, then $n = 2k$ for some integer k. Then $n^2 = 4k^2$, which is even.

Candidate

```
\begin{theorem}
Let $n$ be~an integer. If $n$ is~even,
then $n^2$ is~even.
\end{theorem}

\begin{proof}
If $n$ is~even, then $n=2k$ for~some integer $k$.
Then
\[
n^2 = (2k)^2 = 4k^2,
\]
which is~even.
\end{proof}
```

Example

2. Informal mathematics to a lemma

Prompt

Suppose f is a homomorphism from a group G to a group H. If g has finite order m, then $f(g)$ has order dividing m.

Candidate

```
\begin{lemma}
Let $f\colon G\to~H$ be~a~group homomorphism.
If $g\in G$ has finite order $m$, then the order
of $f(g)$ divides $m$.
\end{lemma}

\begin{proof}
Since $g^m=e_G$, we have
\[
f(g)^m = f(g^m)=f(e_G)=e_H.
\]
Hence the order of~$f(g)$ divides $m$.
\end{proof}
```

Example

3. Mixed prose and formulas

Prompt

The quadratic formula says that the solutions of $ax^2 + bx + c = 0$ are given by $x = (-b \pm \sqrt{b^2 - 4ac})/(2a)$.

Candidate

```
The quadratic formula states that the solutions of
\[
ax^2+bx+c=0
\]
are given by
\[
x=\frac{-b\pm\sqrt{b^2-4ac}}{2a}.
\]
```

The output will rarely be perfect. Inline math markers may need adjustment. Long formulas may need to be split. The structure of theorems and proofs may not match my preferences. Even so, the conversion often saves substantial time, especially when the input is messy or inconsistent.

16.1.2 Imposing Structure on Informal Notes

Even when the source is already in LaTeX, it may be informal. Arguments may be buried inside long paragraphs. Remarks may deserve to be promoted to definitions or theorems. Proofs may be sketches rather than complete presentations. ChatGPT can help impose structure and separate definitions, statements, and proofs cleanly.

Example

Prompt

> Informal text, rewrite:
> Let R be a ring. An element u is a unit if there exists an element v such that $uv = 1$. In a ring, units are important because they are the elements that have multiplicative inverses. If u is a unit, then its inverse is unique.

Candidate

```
\begin{definition}
Let $R$ be~a~ring.
An element $u\in R$ is~called a~\emph{unit}
if there exists an~element
$v\in R$ such that $uv=1$.
\end{definition}

\begin{theorem}
Let $R$ be~a~ring.
If $u\in R$ is~a~unit, then its inverse is~unique.
\end{theorem}

\begin{proof}
Suppose that $v$ and $w$ are inverses of~$u$,
so that $uv=1$ and $uw=1$. Then
\[
v = v(uw) = (vu)w = 1\cdot w = w,
\]
and hence the inverse of~$u$ is~unique.
\end{proof}
```

16.2 Diagnosing and Explaining LaTeX Errors

Large LaTeX projects generate large log files. When something goes wrong, the error messages can be cryptic. The true source of the problem may be far from the line number reported. One of the simplest uses of ChatGPT is to ask it to read a fragment of a log files. I then ask it to explain what is happening.

16.2.1 Finding the First Real Error

When LaTeX encounters an error, subsequent messages often represent a cascade. The key is to find the *first* genuine error.

A practical pattern is:

1. Copy a portion of the log around the first error message.
2. Paste it into ChatGPT with a prompt such as:

> Here is a fragment of a LaTeX log. Identify the first real error. Explain what it means in plain language. Suggest where I should look in the source file.

ChatGPT will usually point to the core issue. It may be

1. a missing `\end{figure}`,
2. an extra right brace,
3. an incompatible package option.

Example

Prompt

> Here are two fragments from LaTeX log files. For each fragment, identify the first real error. Explain it in plain language. Suggest where I should look in the source file.

Fragment A

```
! Extra }, or~forgotten \endgroup.
\@endfloatbox ...pagefalse \outer@nobreak \egroup
\color@endbox
l.206 \end{figure}
```

(continued)

Fragment B

```
! LaTeX Error: Something's wrong--perhaps a~
missing \item.

l.412 \end{enumerate}

?
! Extra }, or~forgotten \endgroup.
l.430 \end{proof}
```

Candidate

```
Fragment A.
The first real error is:
  "Extra }, or~forgotten \endgroup."

This indicates unbalanced braces.
LaTeX noticed the problem when closing the figure,
but the actual mistake is~almost certainly
inside the figure.
Common causes include an~extra } in~a caption,
a missing { in~a command argument, or~malformed
TikZ code.Inspect the contents of~the figure
environment just before
\end{figure}.

Fragment B.
The first real error is:
  "Something's wrong--perhaps a~missing \item."

This means that LaTeX was processing a~list
environment and encountered material
where an~\item was required.
The later brace error is~a~cascade caused by the
list being left in~an inconsistent state.
You should examine the enumerate environment
immediately before the reported \end{enumerate},
checking for~missing or prematurely terminated
\item commands.
```

16.2.2 Explaining Cryptic Messages

Some LaTeX messages are not obvious even when I know where they come from. For example:

- Missing `\endcsname` inserted.
- Runaway argument?
- Illegal unit of measure (pt inserted).

I can ask ChatGPT to explain such messages in context:

> Explain this error message as if you were talking to a mathematician who knows basic LaTeX but not its internals. What are the usual causes? How can I fix it?

The answer will not replace my experience. It can, however, shorten the learning curve. It also provides a checklist of common causes.

Example

Prompt

> Here is a fragment of a LaTeX log files=:
>
> ```
> Runaway argument?
> {List all nonisomorphic groups of size $6$
> whose diagrams have \ETC.
> ./ChatGPTMath.tex:10764: Paragraph ended
> before \text@command was complete.
> <to be read again>
>                        \par
> l.10764 ...ices of size $6$ whose diagrams
> have ?
> ```
>
> Explain this message in plain language. What are the likely causes? How can I fix it?

Candidate

```
The key message is "Runaway argument?"

This means that LaTeX started reading the argument
of a command but never found where it ended.
As a result, it kept reading
subsequent text as part of the same argument.

In this case, the problem is likely a missing
```

(continued)

```
closing brace } in the argument of a command such
as \section, \caption, or a custom macro.

The log message "Paragraph ended before
\text@command was complete"
confirms this: LaTeX reached a paragraph break
(\par) while still inside a command argument.

You should look just before the reported
line number for:
  - a missing },
  - an unmatched {,
  - or a line break inside a macro argument that
  does not allow it.

Fixing the brace structure will usually eliminate
the error.
```

16.2.3 The Limits of Automated Diagnosis

Of course, ChatGPT is not infallible. Sometimes it will misinterpret a message. Sometimes it will suggest a fix that does not apply to my document. As always, I must retain control. My best practice is to treat its suggestions as hypotheses to be tested. I try the smallest change that could resolve the issue. I then recompile and observe the result.

Example

Prompt

Here is a LaTeX error message:

```
! Undefined control sequence.
l.52 \AddtoHook{cmd/\chapter/before}
{\leavevmode}
```

Explain what this means and how to fix it.

Candidate

```
The message "Undefined control sequence" means
that LaTeX encountered
```

(continued)

```
a command it does not know.

In this case, \AddtoHook is provided by the
LaTeX kernel only in
recent releases. If your system uses an older
LaTeX format, the command will be undefined.

A common fix is to update the LaTeX distribution.
Alternatively, you can replace \AddtoHook
with older hook mechanisms or package-
specific solutions.

However, if the document must compile
on an older system, updating may not be possible.
In that case, the suggested fix does not apply,
and the code must be rewritten to avoid
\AddtoHook.
```

16.3 I HATE `log` Files

I am working on a math paper. When I try to typeset it, I get an error message in the `log` file. Sometimes the error message is cryptic. I do not know what to do. ChatGPT can help. Copy the last **two** paragraphs before the error together with the error message. ChatGPT can then propose a line of attack.

If there are no errors, your paper will typeset. You may still need to consult the `log` file for corrections. The `log` file records many kinds of information. Here are some of the most useful.

- **Engine and format information.** At the top of the log, we see which engine and format were used. For example,
  `pdfTeX, Version 3.141592653-2.6-1.40.27`
  `(TeX Live 2025)` together with the LaTeX 2_ε date and patch level.
- **All classes, packages, and files that are loaded.** Every time LaTeX opens a class, package, font definition, or graphics file, the log lists it as a parenthesized file name. Typical entries include `(amsart.cls)` and `(amsmath.sty)`. User files also appear, such as `(newdraw.sty)`. This makes it possible to check exactly which modules were used.

```
This is pdfTeX, Version 3.141592653-2.6-1.40.27
(TeX Live 2025)
(preloaded format=pdflatex)
 restricted \write18 enabled.
entering extended mode
```

```
(./ask.tex
LaTeX2e <2024-11-01> patch level 2
L3 programming layer <2025-01-18>
(/usr/local/texlive/2025/texmf-dist/tex/latex
/amscls/amsart.cls
Document Class: amsart 2020/05/29 v2.20.6
(/usr/local/texlive/2025/texmf-dist/tex/latex
/amsmath/amsmath.sty
```

- **Diagnostic messages about layout quality.** The log lists overfull and underfull boxes. For example,
  `Overfull \hbox(5.9pt too wide) in paragraph at lines 264-264.` A short typewriter fragment shows the offending line. The usual causes are long URLs and long inline code. These messages do not stop the run. They warn us about lines that extend into the margin or have poor spacing.
- **Warnings from LaTeX and packages.** Typical examples include undefined references, unresolved citations, and package-specific issues. For instance, the log may contain
  `LaTeX Warning: Reference 'S:editing' on page 13 undefined`
  or warnings issued by packages such as hyperref. Other common sources include microtype and babel.
- **Page shipping and font usage.** As pages are produced, the log records them as `[1] [2] [3]`, and so on. Near the end of the log, there is a summary of which font files were actually used and embedded in the PDF. You may see entries such as `lmr10.pfb`. You may also see bold or italic variants.
- **A final summary line.** Assuming there was no fatal error, the log usually ends with a line of the form `Output written on ask.pdf (N pages, M bytes)`.

But how do we search these gigantic files for useful information? Ask ChatGPT.

Example

Prompt

> Here is a large fragment from a LaTeX log files. It is hundreds of lines long. How should I search it to find the most useful information? What patterns should I look for first?

Candidate

```
When searching a large LaTeX log files,
focus on structure rather than reading
line by line.
```

(continued)

```
Start by searching for the first line beginning
with an exclamation mark '!'.
This marks the first error reported by LaTeX
and is usually the most important.

Next, search upward for:
  - "Runaway argument?"
  - "Undefined control sequence"
  - "Missing }"
  - "Extra }"
  - "Emergency stop"

Then search for the line numbers reported
after 'l.'. These point to where LaTeX detected
the problem, not necessarily where it was caused,
but they narrow the search area.

Ignore most warnings until all errors are
resolved. Warnings often disappear once
the first real error is fixed.

In short:
  1. Find the first '!'.
  2. Read 10--20 lines above it.
  3. Check the corresponding region in the source.
```

16.3.1 Explaining a Style File Line by Line

Consider a small package such as `mytheorems.sty`, which defines colored theorem environments and their unnumbered variants. I can paste the contents of the file into ChatGPT and ask for an explanation:

Example

Prompt

> Explain this style file line by line. For each macro or setting, describe what it does. Explain how it affects theorem environments.

A good answer will clarify the roles of `\NeedsTeXFormat` and `\ProvidesPackage`. It will also explain `\RequirePackage` and `\newtheoremstyle`. It should then connect these mechanisms to the choices of fonts, colors, and spacing. This is useful when I revisit my own old code. It is also useful when I document it for collaborators and students.

16.3.2 Designing New Theorem Styles

I can also use ChatGPT to design new theorem environments from scratch. For example:

Example

Prompt

> Write a small package that defines theorem environments with italic bodies and colored headings. Use purple for theorems and corollaries. Use blue for lemmas and propositions. Include unnumbered versions. Explain the meaning of each argument to `\newtheoremstyle`.

The result will be a complete `.sty` file, together with a minimal example document showing how to use it. I can then adjust colors. I can also adjust spacing and naming conventions to match my preferences.

Example

Prompt

> Design a small LaTeX package that defines colored theorem styles. Use purple headings for theorems and corollaries, blue headings for lemmas and propositions. The body text should be italic. Provide numbered and unnumbered versions. Explain each argument of `\newtheoremstyle`. Include a minimal example document using results about groups and abelian groups.

Candidate

```
% theoremcolors.sty
\ProvidesPackage{theoremcolors}

\RequirePackage{xcolor}
\RequirePackage{amsthm}
```

(continued)

```
% Purple theorem style
\newtheoremstyle{purplethm}%
  {6pt}   % Space above
  {6pt}   % Space below
  {\itshape} % Body font
  {}      % Indent amount
  {\color{purple}\bfseries} % Head font
  {.}     % Punctuation after head
  {0.5em} % Space after head
  {}      % Head specification

% Blue theorem style
\newtheoremstyle{bluethm}%
  {6pt}
  {6pt}
  {\itshape}
  {}
  {\color{blue}\bfseries}
  {.}
  {0.5em}
  {}

\theoremstyle{purplethm}
\newtheorem{theorem}{Theorem}
\newtheorem{corollary}{Corollary}

\theoremstyle{bluethm}
\newtheorem{lemma}{Lemma}
\newtheorem{proposition}{Proposition}

% Unnumbered versions
\newtheorem*{theorem*}{Theorem}
\newtheorem*{corollary*}{Corollary}
\newtheorem*{lemma*}{Lemma}
\newtheorem*{proposition*}{Proposition}
```

Minimal example:

```
\documentclass{article}
\usepackage{theoremcolors}

\begin{document}
```

(continued)

```
\begin{theorem}
Every subgroup of~an abelian group is~normal.
\end{theorem}

\begin{corollary}
Quotients of~abelian groups are abelian.
\end{corollary}

\begin{lemma}
The identity element of~a~group is~unique.
\end{lemma}

\begin{proposition}
Let $G$ be~a~group and $H\le G$.
If $G$ is~abelian, then $gH = Hg$ for~all $g\in G$.
\end{proposition}

\end{document}
```

16.4 Drawing Diagrams and Managing Graphics

Diagrams occupy a special place in my mathematical writing. They combine conceptual content with visual layout. They often require several rounds of adjustment. ChatGPT cannot see my compiled PDF. It can, however, help with the code that produces it.

16.4.1 From Verbal Description to TikZ Code

As I mentioned in the introduction, ChatGPT can turn a verbal description of an ordered set into TikZ (pronounced like the plural of "tick") code. A typical prompt is:

Example

Prompt

Write TikZ code for the diagram of the ordered set M_3. Use red nodes. Use black edges. Put labels 0 and 1 at the bottom and top. Place the three atoms evenly spaced in the middle row.

(continued)

The resulting code is often close to what I want.

```
\begin{tikzpicture}[scale=1.2]
  % Nodes: bottom (0), three atoms, top (1)
  \node[circle, draw=red, fill=red!20,
     inner sep=1.5pt] (zero) at (0,0) {};
  \node[circle, draw=red, fill=red!20,
     inner sep=1.5pt] (a)    at (-1,1) {};
  \node[circle, draw=red, fill=red!20,
     inner sep=1.5pt] (b)    at (0,1) {};
  \node[circle, draw=red, fill=red!20,
     inner sep=1.5pt] (c)    at (1,1) {};
  \node[circle, draw=red, fill=red!20,
     inner sep=1.5pt] (one)  at (0,2) {};
  % Edges (black)
  \draw[black] (zero) -- (a);
  \draw[black] (zero) -- (b);
  \draw[black] (zero) -- (c);
  \draw[black] (a)    -- (one);
  \draw[black] (b)    -- (one);
  \draw[black] (c)    -- (one);
  % Labels 0 and 1
  \node[below=3pt] at (zero) {$0$};
  \node[above=3pt] at (one)  {$1$};
\end{tikzpicture}
```

This is the diagram we get:

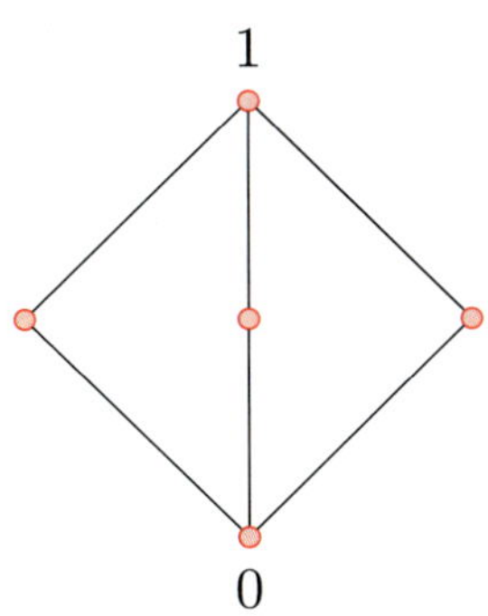

If the geometry is not quite right, I refine it iteratively: moving nodes and adjusting vertical spacing. I then change line styles or colors, add labels, and also remove labels when they distract.

Over time, I can build a small library of such diagrams.

16.4.2 Layout Issues and Font Anomalies

Sometimes a diagram or a block of text is set in the wrong font. A page may unexpectedly switch to sans serif. ChatGPT can help me diagnose such anomalies by inspecting code fragments:

Example

Prompt

In this section, all the prompts appear in sans serif. The last paragraph of Section X also turns sans serif unexpectedly. Here is the LaTeX code around that point. Where is the font being changed? How can I restrict it to the intended region?

By tracking where a font-switching command is opened and not closed, ChatGPT can often point to the missing brace. It may also point to a missing environment end.

Example

Prompt

Here is a fragment of LaTeX source code:

```
\begin{examplebox}
\Prompt
{\sffamily
Give an example of a modular but
nondistributive lattice.
\end{examplebox}

\medskip
This paragraph also appears in sans serif.
```

Why does the sans serif font leak past the prompt? How can I fix it?

Candidate

```
The sans serif font is caused by the command
\sffamily.

It was opened with an explicit brace {
but never closed before the end of the examplebox.
As a result, the font change remains in effect
after the box ends.
```

(continued)

```
The fix is to close the group properly:

{\sffamily
Give an example of a group that is not abelian.
}

Alternatively, the font change can be localized
by using \texttt{\textbackslash textsf\{...\}}.

Once the brace is closed, the surrounding text
returns to the normal serif font.
```

16.5 Glossaries and Notation Lists

Many of my books benefit from a glossary of terms. Many also benefit from a list of notation. ChatGPT can assist with both the selection of entries and the drafting of first versions.

I sometimes ask:

> Here is the source of several chapters. Extract a list of candidate glossary terms. Focus on terms that are explicitly defined. Propose short, student-friendly explanations for each.

Similarly, for notation, I might ask:

> Identify the main symbols introduced in these chapters. Produce a draft list of notation. Include a brief indication of the meaning of each symbol.

The resulting lists are only drafts. They require editing and judgment. Still, they provide a useful starting point and often reveal symbols or terms I had overlooked.

Example

Prompt

> Here is a fragment from two chapters on integration theory. Extract candidate glossary terms and a draft list of notation.

```
Let $f$ be~a~continuous function on~$[a,b]$.
The definite integral $\int_a^b f(x)\,dx$
represents the signed area
under the graph of~$f$ from $a$ to~$b$.
```

(continued)

```
A function $F$ is~an~antiderivative of~$f$
if $F'(x)=f(x)$.
By the Fundamental Theorem of~Calculus,
$\int_a^b f(x)\,dx = F(b)-F(a)$.
```

Candidate

```
Candidate glossary terms:

Definite integral.
For a~continuous function $f$ on~$[a,b]$,
the number $\int_a^b f(x)\,dx$ representing
the signed area under the graph
of $f$ between $a$ and $b$.

Antiderivative.
A function $F$ such that $F'(x)=f(x)$
on~an interval.

Fundamental Theorem of~Calculus.
The theorem stating that if $f$ is~continuous
on~$[a,b]$ and $F$ is~an~antiderivative
of~$f$, then $\int_a^b f(x)\,dx = F(b)-F(a)$.

Draft list of~notation:

$f$ - a~real-valued function.

$[a,b]$ - a~closed interval on~the real line.

$\int_a^b f(x)\,dx$ - the definite integral of~$f$
from $a$ to~$b$.

$F'$ - the derivative of~the function $F$.
```

16.6 Undefined Control Sequences

In this case study, I show how ChatGPT can help interpret a cryptic LaTeX error message and translate it into concrete advice.

16.6.1 Log Fragment

Example

```
./short.tex:34: LaTeX Error: Command \la undefined.

l.34 \renewcommand{\la}
                        {\LaTeX\xspace}
```

Prompt

> Here is a fragment from my LaTeX log:
> `./short.tex:34: LaTeX Error: Command \la undefined.`
> The corresponding line in the source is
> `\renewcommand{\la}{\LaTeX\xspace}`.
> Explain in plain terms what this error means. Should I change this to `\newcommand`, or am I missing a package that defines `\la`?

ChatGPT. The error message "Command `\la` undefined" means that LaTeX has not seen any definition of `\la` before you attempt to redefine it. The command `\renewcommand` is intended for redefining an *existing* macro.

There are two common possibilities.

First, `\la` is your own macro. In that case, it should be introduced with `\newcommand{\la}{...}` rather than `\renewcommand`.

Second, `\la` is expected to be provided by a package. In that case, the package defining `\la` must be loaded earlier in the preamble.

Given your example, it appears that you want a shorthand for the word "LaTeX". In that case, the correct fix is to write

`\newcommand{\la}{\LaTeX\xspace}`

once in the preamble.

16.6.2 Commentary

This answer is essentially what I would tell a student. It explains the distinction between `\newcommand` and `\renewcommand`, lists the two most likely causes of the error, and proposes a reasonable fix.

ChatGPT does not see the entire document, so it cannot know whether `\la` was defined earlier or meant to be supplied by a package. That judgment remains the author's responsibility. Even so, the explanation is valuable: it converts a terse log message into clear reasoning and actionable options, saving time during debugging.

16.7 From Idea to Outline

The first step in a large project is often the hardest. I must decide what the project actually is. I may be aiming at a research paper, a survey, lecture notes, or a book. I must also decide on the intended audience. Only then can I decide which results belong together.

16.7.1 Exploring Possible Scopes

I can use ChatGPT as a sounding board when thinking about the scope of a project. Instead of asking for a single recommendation, I deliberately invite multiple competing perspectives.

Example

Prompt

> I have a series of papers on prime ideals lattices of ring. Here are the main results. Suggest three possible ways to organize them into a single coherent paper. Suggest three possible ways to organize them into a short book.

The responses typically propose different organizing principles. One proposal may group the material by technique. Another may emphasize increasing levels of generality. A third may follow the historical development of the subject. Yet another may be guided by the intended audience.

I may reject most of these suggestions. Even so, the exercise is valuable. By confronting me with alternative structures, ChatGPT forces me to clarify what kind of contribution I want the project to be, and which choices are matters of taste rather than necessity.

16.7.2 Designing a Table of Contents

Once the general scope is clear, I can ask ChatGPT to propose a table of contents. In a paper, this may be little more than a list of sections. In a book, it is usually hierarchical.

Example

Prompt

Here is a list of topics and results that I want to include in a book on [topic].

Propose a table of contents with parts, chapters, and section titles, aimed at beginning graduate students. Briefly indicate the purpose of each part.

The resulting outline is rarely perfect. Some portions are too detailed, while others are too sparse. Even so, it provides a concrete structure to refine. It externalizes choices that were previously implicit.

16.7.3 Iterating on Structure

Structure evolves. As I write, some chapters grow too long, while others turn out to be unnecessary. ChatGPT can help me think about reorganization without taking control of the final shape of the project.

Example

Prompt

Here is my current table of contents.

1. Propose a way to group the chapters into three parts, labeled *basic*, *intermediate*, and *advanced*.
2. Indicate which chapters you would move, and explain why.
3. Suggest where a chapter might be split to improve focus or flow.
4. Point out any chapters whose role seems unclear.

Often I ask a more local question, focusing on a single chapter.

Example

Prompt

This chapter feels crowded. Here are its current section headings.

(continued)

Propose a way to split it into two chapters with clearer and more distinct focus. Briefly describe the role of each resulting chapter in the overall book.

I retain control of the final structure. Still, these suggestions highlight natural groupings and reveal separations I had not previously considered.

16.7.4 Drafting Introductions

ChatGPT is useful for drafting an introduction once the main results are known. At that point, I can ask for a structured draft rather than a free-form paragraph.

Example

Prompt

Write a one-page introduction for a research paper containing the following results.

1. Assume the reader knows basic ring theory but not the specifics of my work.
2. Explain the motivation.
3. State the main results informally.
4. Give a brief overview of the paper's structure.

The draft usually needs revision. Terms may be used too loosely, and emphasis may be uneven. Even so, revising an imperfect introduction is easier than writing one from scratch.

16.7.5 Tailoring Introductions to Different Audiences

I often need multiple introductions for the same work: a conference abstract, a grant summary, or a book chapter introduction. ChatGPT can help translate between these levels.

Example

Prompt

Rewrite the following introduction in two different forms.

(continued)

1. As an abstract of at most 200 words for a general mathematical audience.
2. As a one-paragraph nontechnical summary for a grant reviewer who is not a ring theorist.

The results are rarely perfect, but they often capture the appropriate scale and tone. I can then refine them further.

16.7.6 Conclusions and Open Problems

Similarly, I can ask ChatGPT to help draft a concluding section.

Example

Prompt

Based on the following summary of results:

1. Write a short concluding section highlighting the main contributions.
2. List three to five specific open problems.
3. Suggest directions for further research.

I verify that the proposed open problems are genuine and nontrivial. Nonetheless, the list often provides a useful starting point and stimulates deeper reflection.

16.8 Exercises

Exercise 16.1 Take a TikZ diagram from one of your papers. Ask ChatGPT to simplify the code while preserving the appearance. Compare the original with the simplified version.

Exercise 16.2 Copy a short fragment of a LaTeX log files containing an error you do not immediately recognize. Ask ChatGPT to explain the error and to suggest likely locations in the source. Note whether its suggestions helped you find the actual bug.

Exercise 16.3 Describe in words an ordered set diagram you would like to have. For example, you might specify "a modular lattice of length 4 with exactly three coatoms". Ask ChatGPT to propose TikZ code for the diagram. Decide which parts you would keep and which parts you would edit by hand.

Exercise 16.4 Ask ChatGPT to design a small LaTeX style file snippet, for instance a theorem style with colored headings. Walk through the code line by line and add explanatory comments. This serves as a convenience and also as a mini-tutorial in LaTeX programming.

Exercise 16.5 Find a warning or error message in a LaTeX log files that refers to a line number far from the actual problem. Ask ChatGPT to explain why LaTeX reports the error where it does and to suggest what kinds of source constructs typically cause such behavior.

16.9 Transition

This chapter has focused on the tools that support mathematical writing—LaTeX, diagrams, and the mechanics of turning ideas into reliable documents.

These tools increase efficiency, but they also introduce new points of failure. In the next part, I turn to the limits of ChatGPT in mathematics: where these methods break down, why verification remains essential, and which responsibilities cannot be delegated.

Part IV

Limits and Outlook

17 Limits of ChatGPT for Mathematics

ChatGPT is a powerful assistant for mathematical writing. It is also a useful assistant for exploration, computations, drafting proofs, generating examples, drawing diagrams, and analyzing definitions. But a language model is not a mathematician. ChatGPT predicts the next token. It does not manipulate structures, verify implications, discover the hidden reason why a theorem is true. In this chapter, I describe the fundamental limitations of ChatGPT in mathematics. I also describe how to work safely within those limits.

Many of the failures discussed in this chapter arise when the standard workflow introduced in the Introduction is ignored or abbreviated.

17.1 General Failures

17.1.1 No Formal Reasoning Engine

ChatGPT does not carry out formal reasoning. It does not:

- check that a map is a homomorphism,
- verify distributivity in a lattice,
- prove associativity from first principles,
- or validate a chain of implications.

Instead, it generates text that resembles correct reasoning. When the example is simple, the text often *is* correct. As complexity grows, the model becomes less reliable unless guided by explicit verification steps (see Chapter 3).

G. Grätzer, *ChatGPT for Math*, https://doi.org/10.1007/978-3-032-27286-7_17

17.1.2 Hallucinations

A "hallucination" is a confident but incorrect statement produced by a language model. In mathematics, hallucinations often take specific forms:

- nonexistent theorems (for example, "Birkhoff's lemma on σ-modular ordered sets"),
- false equivalences,
- incorrect citations,
- fabricated examples,
- or proofs that skip a critical case.

Hallucinations are not bugs. They are a natural consequence of predictive text generation. A good workflow recognizes this. It uses ChatGPT's strengths. It does not treat it as an oracle.

17.1.3 Invented Sources and Citations

ChatGPT often fabricates references, journals that do not exist, page numbers that are incorrect, and papers never written by the stated authors.

For this reason, citations must be verified manually. When producing a bibliography, I ask ChatGPT to draft entries. I then confirm publication details through reliable databases. I also consult original sources when needed.

17.1.4 Global Inconsistency

ChatGPT lacks a persistent internal representation of a manuscript. If asked the same question twice, it may change notation, reinterpret definitions, or contradict earlier explanations.

This is not human forgetfulness. It is a property of text generation. To mitigate it, I supply context deliberately:

> *Here is our notation so far...* USE IT CONSISTENTLY IN THE FOLLOWING PROOF.

Consistency improves dramatically once the relevant context is explicitly given.

17.2 Proof Failures

17.2.1 Long Proofs

Long proofs require a sustained chain of dependencies. Definitions must be kept in view. Hypotheses must be tracked. Lemmas and notations must remain stable across many paragraphs. ChatGPT struggles with:

- long-range dependencies,
- subtle hypotheses,
- deferred lemmas,
- and symbols introduced far earlier in the manuscript.

As a result, long proofs produced by ChatGPT often show the same failure modes:

- skipped cases,
- use of unproved results,
- or contradictions with earlier conclusions.

For long arguments, ChatGPT is most useful as a structural assistant. It can outline a proof. It can also act as a critic. I do not rely on it to write the argument itself.

17.3 Trouble with Logic

17.3.1 Symbolic and Computational Limitations

Although ChatGPT can handle small computations, it struggles with:

- long derivations,
- complicated symbolic expressions,
- algebraic simplification at scale,
- and verifying multistep computations.

For symbolic work, a computer algebra system is far more reliable. A short Python or SageMath script is also far more reliable. ChatGPT is excellent at writing such scripts. It is not reliable at executing them.

17.3.2 Unreliable with Hidden Quantifiers

Mathematical definitions often involve nested quantifiers:

$$\forall x\, \exists y\, \forall z\ \Phi(x, y, z).$$

ChatGPT is prone to misreading quantifiers. It may collapse them. It may also shift them implicitly. These slips lead to common errors:

- incorrect generalizations,
- omitted cases,
- or circular reasoning.

A safe approach is to ask ChatGPT to restate a definition with quantifiers before using it in a proof.

17.3.3 Pathological Cases

Humans specialize in finding pathological examples. ChatGPT does not. This difference shows up clearly in the study of differential equations:

- equations with no solutions,
- equations with nonunique solutions,
- solutions that blow up in finite time,
- or equations whose solutions fail to depend continuously on initial conditions.

ChatGPT tends to propose "nice" structures unless asked explicitly for a boundary case. To overcome this, I ask:

Give me the smallest counterexample. Be explicit. Check every operation.

17.4 Weakness with Deep Structure

Some mathematical arguments depend on deep structure:

- subtle structural phenomena,
- hidden symmetries,
- delicate induction hypotheses,
- or distinctions between algebraic levels.

ChatGPT is weakest when the structure is essential but invisible at the surface level. It often fails to identify the turning point of an argument. It misses the key observation on which everything rests.

Because ChatGPT cannot "see" the deep structure, it can produce surface-level arguments. They look plausible. They do not penetrate the essential insight.

17.5 No Mathematical Intuition

Mathematical intuition arises from experience. It grows from many examples. It also grows from pattern recognition guided by structure. It includes a sense of what is plausible. ChatGPT does not possess intuition. It models patterns in text. Its apparent "insights" come from statistical associations. They do not come from conceptual understanding.

This difference becomes clear in research. ChatGPT can assist with exploration (Chapter 14). It cannot replace the instinct needed to choose which paths to follow.

> **Principle**
> Verification can reduce risk, but it cannot transfer responsibility. A mathematically correct-looking argument may pass local checks and still be wrong in structure, scope, or interpretation. Language models can generate errors that are consistent across multiple reformulations, making them resistant to superficial verification. The responsibility for correctness, relevance, and meaning remains with the mathematician, regardless of how carefully the workflow is followed.

17.6 Best Practices for Safe Use

To work safely with ChatGPT in mathematics:

- **Verify everything**, especially proofs and examples.
- Use **redundant prompts** to cross-check results.
- Request **explicit computations** for small cases.
- Provide **clear context** and notation.
- Ask ChatGPT to **criticize its own output**.
- Treat confident claims with caution.

These practices transform ChatGPT into a reliable assistant. They reduce the risk of silent error.

17.7 The Mathematician's Role

ChatGPT accelerates drafting. It also accelerates illustrating. It can accelerate exploration. It can also accelerate editing. The mathematician remains responsible for:

- correctness,
- insight,
- structure,
- exposition,
- and the final product.

The central theme of this book is that ChatGPT extends our capabilities. It does not replace them. We bring intuition. We also bring judgment and mathematical depth. ChatGPT brings speed. It also brings fluency. It can explore many directions at once. Together, the collaboration is powerful, provided we understand the limits.

17.8 Exercises

Exercise 17.1 Ask ChatGPT to state a mathematical theorem that it believes to be true but is actually false. Analyze the statement carefully and identify the precise point where plausibility replaces correctness.

Exercise 17.2 Provide ChatGPT with a correct mathematical statement and ask it to produce a proof. Then ask it to critique its own proof. Compare the critique with your own analysis and note which weaknesses it detects and which it misses.

Exercise 17.3 Ask ChatGPT to generate a convincing argument that relies on an unstated hypothesis. Locate the hidden assumption and rewrite the statement so that the proof would become valid.

Exercise 17.4 Select a task for which ChatGPT performs well and one for which it performs poorly. Ask ChatGPT to explain the difference. Evaluate whether its explanation reflects actual structural limitations of language models.

Exercise 17.5 Ask ChatGPT to list situations in which a mathematician should not use a language model. Compare the list with your own experience and add at least one risk not mentioned.

Exercise 17.6 Review one example earlier in the book where ChatGPT was helpful but potentially misleading. Explain how verification prevented an error and what would have gone wrong without it.

17.9 Transition

This chapter has examined the limits of ChatGPT and the points at which mathematical judgment cannot be delegated.

In the final chapter, I step back from specific techniques to reflect on practice, responsibility, and the long-term role of such tools in the life of a mathematician.

Reflections and Future Directions 18

Throughout this book I have treated ChatGPT as one more tool on my desk. It sits next to LaTeX, TikZ, and a computer algebra system. I also keep scrap paper nearby for calculations. I have seen ChatGPT assist me in many settings to help check and polish proofs, propose examples and counterexamples. It can support computation and experimentation, explain concepts at multiple levels, design courses and exercises, smooth the mechanics of LaTeX and diagrams, and support large writing projects.

In this final chapter, I want to step back from the details. I want to reflect on the broader picture. What changes when I work alongside a system like ChatGPT? What responsibilities remain unchanged? Some may even become more pressing. How might my habits of proof and exposition change? How will my teaching evolve in the years ahead?

18.1 What I Have Learned

Let me begin by summarizing what the previous chapters show. I do this in a more conceptual way.

18.1.1 Language as a Mathematical Tool

ChatGPT is a system that works exclusively with language. It does not manipulate formal proofs internally. It does not maintain a database of theorems. A computer algebra system maintains identities and algorithms. ChatGPT does something different. It has learned patterns of use. It has learned how mathematicians write proofs, explain concepts, and typeset formulas. It has even learned how they phrase emails to students and coauthors.

From this fact, several consequences follow.

G. Grätzer, *ChatGPT for Math*, https://doi.org/10.1007/978-3-032-27286-7_18

- ChatGPT is particularly good at tasks that involve *rephrasing*. It can rewrite a proof. It can reorganize an explanation. It can also change the level of detail. It can translate between languages. It can adapt an argument for a different audience.
- It is good at tasks that involve *structure in text*. It can construct outlines. It can propose tables of contents. It can also help design exercise sets. It can suggest lecture plans.
- It is moderately good when language and simple structure meet. It can describe small finite structures and candidate examples. It writes short pieces of code very well.

These strengths explain why so many of the useful workflows in this book combine text with structure. They also draw on modest combinatorics.

18.1.2 Assistance, Not Authority

At the same time, ChatGPT is unreliable as an authority. ChatGPT can make mistakes in definitions, omit hypotheses, present proofs that look convincing but are incorrect. It can also invent references and historical anecdotes.

It really does not know when it is making a mistake.

For me as a mathematician, this is decisive. The central value of my subject is correctness. I want to distinguish true statements from false statements by means of proofs that can be checked line by line. A tool that cannot guarantee correctness cannot be trusted to make mathematical decisions on my behalf.

This leads to a guiding principle that I have repeated throughout the book:

ChatGPT is an *assistant*, not an oracle.

Used as an assistant, it can lighten the load. It can also make some tasks more enjoyable. Treated as an oracle, it is dangerous.

18.2 Benefits to My Practice

Within that constraint, the benefits are substantial. It is helpful to group them by aspect of practice.

18.2.1 Proofs and Examples

In Chapter 5, I showed how ChatGPT can

1. reorganize a proof,
2. make its logical structure explicit,

3. suggest where assumptions are used,
4. propose alternative proofs in different styles,
5. highlight different aspects of the underlying structure,
6. generate candidate examples of small finite structures,
7. diagnose flawed arguments,
8. point to suspicious steps and missing cases.

None of this replaces my judgment. Every proof and example must still be checked. But ChatGPT can act as a persistent interlocutor. It can be reasonably knowledgeable and never tires of rewriting a paragraph.

18.2.2 Explanation and Teaching

In Chapters 9 and 15, I considered explanation and teaching. Here the strengths of a language model are especially clear. It can

- generate multi-level explanations of the same concept. One version can be aimed at undergraduates. Another can be aimed at graduate students. A third can be aimed at experts in other fields. I can then refine what it produces.
- help rewrite definitions and theorems as lecture notes. It can also help turn them into blackboard plans. It can draft handouts with informal motivation and examples.
- generate exercises and problem sets. It can draft exam questions in graded collections. It can also draft solutions and hints for my own use as an instructor.
- help me create mock oral exams and question banks. These can help students practice articulating mathematics aloud.

I remain responsible for mathematical content, for alignment with course goals, and academic integrity. Even so, many repetitive and time-consuming aspects of teaching is facilitated by ChatGPT.

18.2.3 LaTeX, Diagrams, and Writing

Chapters 16 addressed the mechanical side of writing. They showed how ChatGPT can help

- convert messy text into structured LaTeX,
- diagnose error messages in LaTeX log files and explain them in plain language,
- design style files, document them, and assist with theorem environments,
- generate TikZ code for diagrams from verbal descriptions,
- draft shell scripts for graphics handling and draft small automations for compilation workflows,

- plan and draft large documents, propose a table of content, and draft introductions,
- clean bibliographies and help standardize them,
- support indexing and glossary work.

These are not glamorous tasks, but they are essential. By reducing friction, ChatGPT allows me to devote more time and attention to the mathematics itself.

18.3 Risks and Responsibilities

The benefits come with risks. Some are obvious. Others emerge more slowly.

18.3.1 Overreliance and Loss of Skill

One risk is overreliance. If I outsource too much to an automated assistant, I may gradually lose skills. I must know how to write a clean proof without help and debug a LaTeX error by reading the log.

The remedy is not to avoid ChatGPT. The remedy is to use it consciously. For tasks at which I want to remain skilled, it can be wise to alternate between assisted and unassisted work. It can also be wise to use ChatGPT only at later stages of editing. For my students, I may do some assignments without AI assistance, while I may allow AI assistance in other assignments.

18.3.2 Hallucinations and Subtle Errors

Another risk is the phenomenon now commonly called "hallucination". This is the confident production of false statements. In mathematics these errors can take several forms:

- proofs that overlook a special case,
- proofs that misuse a lemma,
- examples that do not satisfy all stated properties,
- invented theorems,
- misattributed results,
- incorrect historical and biographical claims.

For my own practice, the rule must be absolute. No statement generated by ChatGPT is to be accepted as correct without independent verification. This may sound obvious. The fluency and confidence of the system can make it easy to forget.

18.3.3 Attribution and Authorship

There are also questions of attribution. If ChatGPT contributes significantly to the wording of a proof, what should be acknowledged? If it contributes to the structure of a chapter, what should be acknowledged? Practices are still evolving. Two principles seem reasonable to me:

- Mathematical credit is reserved for human beings who take responsibility for content.
- Substantial use of AI tools in producing text may be mentioned in acknowledgments. It may also be mentioned in a preface. This is most natural in books and expository articles.

Different journals and publishers are beginning to articulate their own policies. I will need to adapt to these. I may also help shape them.

18.4 ChatGPT and Formal Methods

In this book, I have not focused on the intersection between language models and formal proof assistants: Coq, Lean, and Isabelle. This area is developing rapidly. It has its own literature. Nonetheless, a few remarks may be useful.

18.4.1 Different Strengths

Formal proof systems provide certainty. They do so at a cost. They work in a restricted language. They require detailed justification of every step. This can be tedious. Language models provide flexibility and speed in ordinary mathematical language. They also do so at a cost. Their output is unreliable.

It is easy to imagine workflows that combine these tools. A language model can propose a proof sketch. A proof assistant can then force full detail. A proof assistant can also find a gap. A language model can help translate the gap into something I can repair. Experiments of this kind are already under way.

18.4.2 My Role as a Mathematician

In such workflows, my role may shift. I may focus more on selecting definitions. I may also focus more on conjecturing statements. I may guide the overall direction of research. I then interpret the results. This is not entirely new. Mathematicians using computer algebra systems have long played similar roles. The novelty is the breadth of tasks now assisted by language-based tools.

It is too early to say how far this will go. What seems clear is that mathematical understanding will remain a human responsibility. I include my own work in that. I still want to see why a theorem is true. I still want to see how it fits into a larger picture. I also want to see what it suggests next.

18.5 My Personal Practice and Habits

Each mathematician will develop a personal style of working with tools like ChatGPT. Nonetheless, a few habits suggest themselves for my own work.

18.5.1 Separate "Thinking" from "Polishing"

ChatGPT can be useful to separate stages of work. In an early stage, I may write proofs and notes on my own. I may even write by hand. This keeps a direct connection with the material. In later stages, I can invite ChatGPT to help reorganize. I can also ask it to clarify. I can then ask it to polish.

This separation reduces the risk that the model's suggestions will shape my thinking too early. It also preserves the satisfaction of solving problems without external help.

18.5.2 Keep a Prompt Notebook

Over time I will discover prompts and workflows that work well for me. I may find a good way to ask for an alternative proof. I may also find a good way to ask for a diagram. I may find a good way to ask for an exercise set. I may also find a good way to ask for a log-file explanation. It is worthwhile to keep a small "prompt notebook". It can be a file of examples. I can reuse it and also adapt it.

This notebook is not a fetish. It captures experience. It plays a role analogous to that of a macro file. It also resembles a collection of commonly used lemmas.

18.6 For Younger Mathematicians

You may be be a student or a younger mathematician. You have grown up with tools like ChatGPT. Some cautions in earlier chapters may feel obvious. Other cautions may seem old-fashioned. Still, it is worth making a few points explicit.

18.6.1 Learn to Do Things by Hand

Even if you plan to use ChatGPT regularly, you should learn how to do core tasks without it, read and write proofs, and construct examples. Make sure you can debug basic LaTeX problems and plan a short paper or talk.

These skills form the basis of professional competence. Once they are solid, tools can amplify them.

18.6.2 Be Honest About Your Use of Tools

In coursework and examinations, follow your instructor's rules about AI assistance. In research, be honest with yourself and others. Be clear about what you have done and what the tool has done. If a proof owes its structure to ChatGPT, you may decide no explicit mention is needed. That decision presumes you have checked and understood every step. If a passage of text is largely written by the tool, a brief acknowledgment may be appropriate.

18.6.3 Cultivate Taste and Judgment

Tools can produce many things. They can produce proofs, examples, explanations, and diagrams. Not all of these outputs are good. Mathematical maturity includes taste and judgment. You learn which proofs are illuminating, natural, instructive, and helpful for a particular audience. No tool can do this for you.

18.7 Looking Ahead

It is tempting to speculate about dramatic future developments in a concluding chapter. One could imagine fully automated theorem provers that collaborate with language models, systems that propose conjectures beyond our current imagination, and assistants that read the entire mathematical literature. Some of these may come. Others may remain science fiction.

From the vantage point of this book, a more modest prediction seems appropriate. For the foreseeable future, systems like ChatGPT will remain powerful assistants. They will also remain imperfect but more integrated into my tools. They will still surprise me in ways I cannot now predict.

What is unlikely to change is the core of mathematical practice. I work with definitions, conjectures, and explanations. These activities may be assisted and accelerated by machines. They will still require human creativity and judgment. Even if a computer produces a proof, I will still want understanding and why that theorem is true.

18.8 A Final Word

When I first began to use ChatGPT, I did not expect it to affect my mathematical work. It seemed like a convenient tool for emails, for summarizing documents, and answering general questions. Over time, it became clear that much of what I do as a mathematician has a linguistic and structural side. This includes proving theorems, drawing diagrams, and writing papers. It also includes teaching students and communicating with the world. A system like ChatGPT can support this side of the work.

In writing this book, I have tried to document helpful workflows. I have also tried to be honest about limitations and risks. My hope is that you will treat ChatGPT neither as a toy nor as a mysterious oracle. Treat it as a practical assistant. It can make some aspects of mathematical life easier and more pleasant. It leaves the core of our activity in human hands. This core is understanding and creating mathematics.

18.8.1 What Changes: And What Does Not

ChatGPT changes how quickly I can explore ideas, draft alternatives, and test small cases. It does not change what counts as understanding, what counts as proof, or who is responsible for correctness. Those remain human obligations. Used well, ChatGPT shortens the path from question to insight. Used carelessly, it shortens the path to error.

ChatGPT is an *assistant*, not an oracle. Used in that spirit, it can be a remarkably good companion for the working mathematician.

18.9 Exercises

Exercise 18.1 Review how you currently use ChatGPT (or similar tools) in your mathematical work. List the tasks for which it has been genuinely helpful and the tasks for which it has been unreliable or distracting.

Exercise 18.2 Ask ChatGPT to summarize the main principles of safe use that emerge from this book. Compare its summary with your own understanding and note any principles it omits or misrepresents.

Exercise 18.3 Choose one habit or workflow described earlier in the book that you intend to adopt permanently. Explain why it improves your mathematical practice and how you will integrate it into your routine.

Exercise 18.4 Ask ChatGPT how it expects language models to influence mathematical writing over the next decade. Evaluate which predictions are plausible and which rest on unsupported assumptions.

Exercise 18.5 Reflect on a situation in which overreliance on automation could weaken mathematical judgment. Describe how you would redesign the workflow to keep the mathematician firmly in control.

Exercise 18.6 Write a short paragraph, for yourself, summarizing your personal stance on using AI tools in mathematics: what you will use them for, what you will avoid, and where responsibility must always remain human.

Glossary

A

Alternative proof A proof of a statement that uses a different method, viewpoint, or structure than a previously given proof, often revealing additional insight or dependencies.

Assistant, not an oracle The guiding principle that ChatGPT can support mathematical work by drafting, organizing, and suggesting, but cannot certify correctness or replace mathematical judgment.

Boundary case An extreme or limiting case of a definition or statement that often exposes hidden assumptions or failure modes.

Brute-force checking Verification by explicit enumeration and computation of all cases in a finite setting.

Candidate example An example proposed for a given purpose that has not yet been fully verified.

Commutative diagram A diagram representing objects and morphisms in which all directed paths with the same endpoints represent equal compositions.

Context window The finite portion of prior text that a language model uses when generating a response.

Counterexample A mathematical object that satisfies the hypotheses of a statement but violates its conclusion.

Diagrammatic reasoning The use of diagrams to expose structure, relationships, or constraints that may be less visible in symbolic form.

Experimental mathematics The use of computation, enumeration, or simulation to explore patterns, test conjectures, or guide intuition.

Failure mode A characteristic way in which ChatGPT produces incorrect, misleading, or incomplete mathematical output.

Finite enumeration The systematic listing of all objects of a given type within a finite universe.

Flow chart A directed diagram representing a procedural or logical workflow with branching and merging paths.

Hallucination A confident but incorrect statement produced by a language model, such as a false theorem, incorrect example, or invented reference.

G. Grätzer, *ChatGPT for Math*, https://doi.org/10.1007/978-3-032-27286-7

Hidden assumption An unstated condition that is implicitly used in an argument or explanation.

Human in the loop A workflow in which a human mathematician actively verifies, corrects, and interprets all AI-generated output.

Index entry candidate A proposed term or phrase identified as potentially deserving inclusion in the index.

Informal explanation An explanation intended to convey intuition or motivation rather than formal proof.

Large Language Model (LLM) A predictive text system trained on large corpora that generates responses by modeling statistical patterns in language.

Localizing errors The process of identifying the specific step, hypothesis, or construction where an argument or computation fails.

Log file The diagnostic output produced by LaTeX, containing errors, warnings, and compilation information.

Mechanical verification Verification carried out by explicit computation, case checking, or automated testing rather than abstract proof.

Minimal counterexample A counterexample that is minimal with respect to size or complexity, often clarifying why a statement fails.

Minimal working example (MWE) A smallest self-contained LaTeX document that reproduces a specific error or behavior.

Model confidence vs. correctness The distinction between fluent presentation of an answer and its actual mathematical validity.

Multi-level explanation Parallel explanations of the same concept tailored to different levels of background or expertise.

Nonexample An object that deliberately fails to satisfy a definition, used to clarify the scope of a concept.

Ordered set diagram A diagram representing an ordered set by placing elements at different levels and drawing covering relations.

Pathological case An example that behaves unusually or violates naive expectations, often revealing the limits of a definition or theorem.

Pattern recognition The identification of regularities or trends in computed or enumerated data.

Plausible but incorrect proof A proof that follows familiar rhetorical patterns but contains a logical gap, missing hypothesis, or invalid inference.

Predictive text model A model that generates text by estimating which tokens are most likely to follow a given context.

Proof sketch An outline of a proof that indicates the main ideas without supplying full detail.

Redundant prompting The practice of asking the same mathematical question in multiple ways to expose inconsistencies or hidden errors.

Runaway argument A LaTeX error caused by an unclosed command argument, leading LaTeX to read unintended text as part of the argument.

Small-case testing Verification by checking a statement on small finite instances.

Small finite structure A finite mathematical object used for explicit computation, testing, or illustration.

Standard ChatGPT-math workflow A disciplined interaction pattern consisting of goal specification, candidate generation, explicit structuring, verification, and revision.

Token A basic unit of text processed by a language model, such as a word fragment or symbol.

Trust but verify The principle that AI-generated mathematical content must always be independently checked.

Undefined control sequence A LaTeX error indicating the use of a command that has not been defined.

Verification step The stage in a workflow where correctness is checked independently of AI output.

Verified example An example whose properties have been explicitly checked and confirmed.

Visualization as verification The use of diagrams or pictures to test whether claimed structural properties actually hold.

Prompt Patterns for Mathematicians

B

In the main chapters, I presented many concrete workflows in which ChatGPT can assist. It can assist with proofs. It can assist with examples. It can assist with computation. It can assist with teaching. It can assist with LaTeX. It can also assist with writing and with outreach. In this appendix I collect a selection of prompt patterns. I group them by task. They are intended as templates. You can adapt them to your own topics. You can also adapt them to your own notation.

Throughout, replace bracketed phrases such as `[topic]` with your own material. Do the same for placeholders such as `[system]`.

B.1 Proofs and Examples

B.1.1 Polishing a Proof

Prompt P1 (sentence-by-sentence editing).
Here is a proof in LaTeX. Go through it sentence by sentence. For each sentence, print the original. Then propose an improved version for clarity and style. Briefly explain the change. Wait for me to accept ('y') or reject ('n') each change before moving on. Do not alter the mathematical content.

B.1.2 Alternative Proofs

Prompt P2 (different style of proof).
Here is a proof of Lemma X. Suggest an alternative proof that uses [order-theoretic] ideas. If that is not natural, try [combinatorial] ideas. If that is not natural, try [categorical] ideas. Aim the level at beginning graduate students. Explicitly state any additional hypotheses you need.

G. Grätzer, *ChatGPT for Math*, https://doi.org/10.1007/978-3-032-27286-7

B.1.3 Examples and Counterexamples

Prompt P3 (finite example under constraints).
I am looking for a small finite group with properties (A) and (B). I also want property (C). I do not want property (D). Suggest a candidate with at most n elements. Describe it explicitly. For example, present it as a subgroup of a symmetric group. Explain why it has the listed properties. I will check the details myself.

Prompt P4 (testing a conjecture on small cases).
I conjecture that every finite structure with properties (A) and (B) also has property (C). Think about small cases with size at most n. Try to find a counterexample. If you propose a candidate, explain carefully why it satisfies (A). Explain carefully why it satisfies (B). Explain carefully why (C) fails. If you cannot find a counterexample, say so. Explain which cases you have effectively considered.

B.2 Computation and Experimentation

B.2.1 Designing Code

Prompt C1 (from description to code).
I want to run the following experiment: [informal description]. Write short, clear code in [Python]. If you prefer, use [Sage]. You may also use [GAP] or [Mathematica]. Use descriptive variable names. Add comments. Avoid unnecessary optimizations so that I can modify the code easily.

B.2.2 Debugging

Prompt C2 (explain an error).
Here is a code fragment in [system]. Here is the error message it produces. Explain in plain language what is going wrong. Suggest a minimal fix. Do not rewrite the entire program. Focus on this error and the surrounding lines.

B.2.3 From Output to Tables

Prompt C3 (LaTeX table from raw output).
Here is the text output of my computation: [paste lines]. Convert it into a LaTeX table using `\begin{tabular}`. Use one row per value of n. Add column headers for each quantity. Do not invent new data. Restructure only what is given.

B.3 Explaining Math and Teaching

B.3.1 Multi-level Explanations

Prompt E1 (three audiences).

Explain the following definition or theorem: [statement]. Give a version for a strong undergraduate. Give a version for a beginning graduate student in the area. Give a version for an expert in another branch of math. In each version, include at least one concrete example.

B.3.2 Simplifying Without Losing Correctness

Prompt E2 (plain language with safeguards).
Rewrite this explanation in simpler English. Aim at a mathematically literate but nonexpert reader. Avoid jargon where possible, but keep the mathematical claims accurate. When you omit a hypothesis, mark it in brackets so that I can restore it if necessary. Do the same for any technical caveat.

B.3.3 Designing Exercises

Prompt T1 (graded exercise set).
Here is a definition. Here is a basic theorem: [text]. Create ten exercises. Group them into three levels. First, include routine exercises that check understanding. Next, include intermediate exercises that develop technique. Finally, include one or two challenging problems. Aim at strong undergraduates. Provide only the statements, not the solutions.

B.3.4 Solutions and Hints

Prompt T2 (solutions plus hints).
For each of the following exercises, first write a full solution. Aim at the level of an instructor's manual. Then rewrite that solution as a short hint. The hint should be suitable for a student who is stuck. Keep the hints brief. Keep them suggestive rather than explicit.

B.3.5 Course Outlines

Prompt T3 (course plan).
I am planning a [10] week course on [topic] for [audience]. Propose a week-by-week outline with topics. Include key theorems. Include one example per week. You may include a second example when it is helpful. Indicate where a review session might fit. Indicate where a midterm exam might fit.

B.4 LaTeX, Diagrams, and Mechanics

B.4.1 Log Files Diagnosis

Prompt L1 (first real error).
Here is a fragment of a LaTeX log files around an error. Identify the *first* real error. Explain in plain language what it means. Suggest where in the source I should look. If it helps, mention likely causes in a figure environment. You may also mention likely causes in a list environment. Do not just quote the message; interpret it.

B.4.2 TikZ Diagrams from Descriptions

Prompt L2 (diagram in TikZ).
Write TikZ code for the diagram of the ordered set [name or description]. Put nodes at roughly three levels. Draw edges as straight line segments. Put label 0 at the bottom. Put label 1 at the top. Use simple coordinates so that I can adjust the layout by hand. Do not include any preamble. Provide only the TikZ picture environment.

B.4.3 Style Files and Theorem Environments

Prompt L3 (designing a theorem style).
Write a small LaTeX package `mytheorems.sty` that defines theorem environments with italic bodies and colored headings. Use purple for theorems and corollaries. Use blue for lemmas and propositions. Include unnumbered starred versions. Comment each block of code to explain what it does.

B.5 Writing Papers and Books

B.5.1 Introductions and Abstracts

Prompt W1 (paper introduction).
Here is a description of the main results of my paper: [text]. Write a one-page introduction. Aim at readers familiar with [background area]. Explain the motivation. State the results informally. Outline the structure of the paper. Avoid overselling. Keep the tone typical of a research article.

B.5.2 Related Work

Prompt W2 (constrained related-work section).
Using only the following references: [list or BibTeX], write a related-work section. Explain how these papers fit together. Explain how my results (described here: [summary]) differ from them. If my results extend them, explain how. Do not mention any references other than those given.

B.5.3 Public Talks and Popular Articles

Prompt W3 (public lecture outline).
Based on this summary of my work on [topic], propose a 45-minute public lecture outline for a general audience. Include 5–7 sections. Include a motivating opening. Include at least one simple example. Include a concluding section with open questions. Avoid formulas. Focus on ideas and analogies.

Prompt W4 (plain-language summary).
Write a 300-word plain-language summary of this work. Make it suitable for a departmental newsletter. It should also fit a university press release. Avoid technical notation. Define any specialized terms in simple language.

These patterns are not prescriptions. Their purpose is to illustrate how to phrase requests so that ChatGPT can act as a useful assistant. Be clear about the task. Be explicit about the audience. Be honest about the need for human checking. Be equally honest about the need for human editing of all mathematical content.

B.5.4 Explanation

This message usually means that somewhere in the material of the floating environment (here, a `figure`), the grouping of braces has become unbalanced. There may be an extra `}`. There may be a missing `{`. The error is reported at the `\end{figure}` line. The actual problem is almost always *inside* the figure. It may be in the caption. It may be in a command argument. It may be in drawing code.

The mention of `\@endfloatbox` and `\color@endbox` indicates that the error was detected while ending the box for the float. It was not detected in ordinary text.

B.5.5 Typical Fix

I inspect the figure environment around line 206 in `ask.tex`:

```
\begin{figure}
\centerline{\includegraphics[scale=0.8]
          {original.png}}
\caption{xxx}
\end{figure}
```

In this simple example there is no error. In a real case I might find something like:

```
\caption{A diagram of~$\mathsf{M} 3
}   % missing closing $
```

Or I might find:

```
\caption{Compare with Figure~\ref{F:previous}}
```

Here the label `F:previous` may have been misspelled in the corresponding `\label`. An edit can then disturb brace structure by accident.

The remedy is to check every `{` and every `}`. I check math mode delimiters. I also check command arguments. I continue until the groups are balanced.

B.5.6 How ChatGPT Can Help

A useful prompt is:

> Here is a fragment of my LaTeX log files containing an 'Extra } or forgotten \endgroup' error at \end{figure}. Explain what this usually means and where in the source I should look. Suggest common mistakes inside figure environments that could cause this, and how to fix them.

ChatGPT can then remind me of typical culprits. Mismatched math mode delimiters are common. Braces in captions are common. Unbalanced optional arguments can also cause this.

B.6 Case 1: "Undefined Control Sequence"

B.6.1 Log Fragment

```
./short.tex:34: LaTeX Error: Command \la undefined.

l.34 \renewcommand{\la}
                            {\LaTeX\xspace}
```

B.6.2 Explanation

The message "Command `\la` undefined" means that I am trying to use `\la` before it has been defined. In this example, I attempt `\renewcommand{\la}{...}`. But `\renewcommand` is meant for redefining an existing command. If `\la` has never been defined, I must use `\newcommand` instead.

Alternatively, `\la` might be defined in a package that has not yet been loaded.

B.6.3 Typical Fix

If `\la` is my own macro, I change

```
\renewcommand{\la}{\LaTeX\xspace}
```

To

```
\newcommand{\la}{\LaTeX\xspace}
```

Earlier in the preamble. If `\la` should come from a package, I load that package before the line that uses or redefines `\la`.

B.6.4 How ChatGPT Can Help

I can ask:

> My log says 'LaTeX Error: Command \la undefined' at a line where I write \renewcommand{\la}. Explain why this is happening and whether I should change it to \newcommand or load an additional package.

ChatGPT will typically explain the distinction between `\newcommand` and `\renewcommand` and suggest the appropriate change.

B.7 Case 2: "Runaway Argument?" and "Paragraph Ended ..."

B.7.1 Log Fragment

```
Runaway argument?
{This is~the beginning of~a~caption that
never closes properly
! Paragraph ended before \@caption was complete.
<to~be read again>
                       \par
l.119 \end{figure}
```

B.7.2 Explanation

A "Runaway argument?" means that LaTeX started reading the argument of a command and never saw the closing brace. Here the command is `\caption`. LaTeX kept reading until a paragraph ended. It may also stop when a different command forces it to. It then complains that the argument is incomplete.

In practice, this often happens when I comment out a brace. It can also happen when I delete part of a caption.

B.7.3 Typical Fix

I look at the `\caption` line in the figure at or before line 119:

```
\begin{figure}
\centerline{\includegraphics[scale=0.8]{original.png}}
\caption{This is~a~long caption
\end{figure}
```

Here the closing brace } is missing. I add it:

```
\caption{This is~a~long caption}
```

And I recompile. If the caption contains math or nested braces, I check that all of them are balanced.

B.7.4 How ChatGPT Can Help

A good prompt is:

> LaTeX reports a 'Runaway argument?' and 'Paragraph ended before \caption was complete' at the end of a figure environment. Explain in plain terms what this means. List the most likely mistakes in the caption that I should look for.

ChatGPT can then remind me to look for missing `}`. It can also remind me to look for unclosed `$...$`. Nested braces can also be the culprit. A comment marker can create the same effect.

B.8 Case 3: "Illegal Unit of Measure (pt Inserted)"

B.8.1 Log Fragment

```
! Illegal unit of~measure (pt inserted).
<to~be read again>
                         5
l.203 \vspace{5}
```

B.8.2 Explanation

In commands such as `\vspace`, `\hspace`, and `\setlength`, LaTeX expects a *dimension*. A dimension is a number followed by a unit, such as `5pt` or `1cm`. Another common unit is `2em`. Writing `\vspace{5}` is not legal, because LaTeX does not know what unit "5" should have.

When it encounters a bare number, it complains. It then assumes points (`pt`) by default.

B.8.3 Typical Fix

I replace

```
\vspace{5}
```

By

```
\vspace{5pt}
```

Or, if I intended a larger space,

```
\vspace{5mm}
```

I always include a unit.

B.8.4 How ChatGPT Can Help

Prompt:

> LaTeX reports 'Illegal unit of measure (pt inserted)' at a line \vspace{5}. Explain what this error means and how to correct it. Mention other commands where the same mistake would occur.

The answer will typically mention `\vspace` and `\hspace`. It will usually also mention `\setlength` and related commands.

B.9 Case 4: "Label(s) May Have Changed" and Undefined References

B.9.1 Log Fragment

```
LaTeX Warning: Reference 'F:prompt' on~page 2 undefined
on~input line 119.

LaTeX Warning: Label may have changed.
Rerun to~get cross-references right.
```

B.9.2 Explanation

These are warnings, not errors. The first warning says that I used `\ref{F:prompt}` or `\eqref{F:prompt}` somewhere, but no `\label{F:prompt}` was defined. The second warning is the familiar message about cross-references and the table of contents. I must rerun LaTeX so that labels and references are synchronized.

B.9.3 Typical Fix

For the undefined reference:

- I check whether I misspelled the label in the `\label` command.
- I check the spelling in the `\ref` as well.
- If the label corresponds to a figure that I deleted, I remove the reference. I can also update it.

For the “Label(s) may have changed” message, I rerun LaTeX. I do this until the warning disappears. This usually takes one run. Sometimes it takes two. If undefined references remain, I fix them first.

B.9.4 How ChatGPT Can Help

Prompt:

> My log shows ‘Reference ‘F:prompt’ undefined’ and ‘Labels may have changed’. Explain the difference between these warnings and what steps I should take to resolve them. Mention how many times I should rerun LaTeX.

ChatGPT will typically explain that undefined references require fixing the source. It will then explain that changed labels require another compilation.

The cases in this appendix are not exhaustive, but they illustrate a useful pattern. When I encounter a puzzling error or warning:

1. I copy a short fragment of the log around the error.
2. I ask ChatGPT to explain the message in plain language and to suggest likely causes.
3. I inspect the suggested locations in my source myself.
4. I then make the actual fixes.

As always, ChatGPT is an assistant. It can shorten the search. I remain in control of the document.

Index

G. Grätzer, *ChatGPT for Math*, https://doi.org/10.1007/978-3-032-27286-7

W

Z

MIX
Papier | Fördert gute Waldnutzung
FSC® C083411

Zeitfracht Medien GmbH
Ferdinand-Jühlke-Straße 7
99095 Erfurt, Deutschland
produktsicherheit@kolibri360.de